MONOGRAPHS ON STATISTICS AND APPLIED PROBABILITY

General Editors

D.R. Cox, D.V. Hinkley, D. Rubin and B.W. Silverman

1 Stochastic Population Models in Ecology and Epidemiology *M.S. Bartlett* (1960)

2 Queues *D.R. Cox and W.L. Smith* (1961)

3 Monte Carlo Methods *J.M. Hammersley and D.C. Handscomb* (1964)

4 The Statistical Analysis of Series of Events *D.R. Cox and P.A.W. Lewis* (1966)

5 Population Genetics *W.J. Ewens* (1969)

6 Probability, Statistics and Time *M.S. Bartlett* (1975)

7 Statistical Inference *S.D. Silvey* (1975)

8 The Analysis of Contingency Tables *B.S. Everitt* (1977)

9 Multivariate Analysis in Behavioural Research *A.E. Maxwell* (1977)

10 Stochastic Abundance Models *S. Engen* (1978)

11 Some Basic Theory for Statistical Inference *E.J.G. Pitman* (1979)

12 Point Processes *D.R. Cox and V. Isham* (1980)

13 Identification of Outliers *D.M. Hawkins* (1980)

14 Optimal Design *S.D. Silvey* (1980)

15 Finite Mixture Distributions *B.S. Everitt and D.J. Hand* (1981)

16 Classification *A.D. Gordon* (1981)

17 Distribution-free Statistical Methods *J.S. Maritz* (1981)

18 Residuals and Influence in Regression *R.D. Cook and S. Weisberg* (1982)

19 Applications of Queueing Theory *G.F. Newell* (1982)

20 Risk Theory, 3rd edition *R.E. Beard, T. Pentikainen and E. Pesonen* (1984)

21 Analysis of Survival Data *D.R. Cox and D. Oakes* (1984)

22 An Introduction to Latent Variable Models *B.S. Everitt* (1984)

23 Bandit Problems *D.A. Berry and B. Fristedt* (1985)

24 Stochastic Modelling and Control *M.H.A. Davis and R. Vinter* (1985)

25 The Statistical Analysis of Compositional Data *J. Aitchison* (1986)

26 Density Estimation for Statistical and Data Analysis *B.W. Silverman* (1986)

27 Regression Analysis with Applications *G.B. Wetherill* (1986)

28 Sequential Methods in Statistics, 3rd edition *G.B. Wetherill* (1986)

29 Tensor Methods in Statistics *P. McCullagh* (1987)

30 Transformation and Weighting in Regression *R.J. Carroll and D. Ruppert* (1988)

31 Asymptotic Techniques for use in Statistics *O.E. Barndoff-Nielson and D.R. Cox* (1989)

32 Analysis of Binary Data, 2nd edition *D.R. Cox and E.J. Snell* (1989)

33 Analysis of Infectious Disease Data *N.G. Becker* (1989)

34 Design and Analysis of Cross-Over Trials *B. Jones and M.G. Kenward* (1989)

35 Empirical Bayes Method, 2nd edition *J.S. Maritz and T. Lwin* (1989)

36 Symmetric Multivariate and Related Distributions *K.-T. Fang, S. Kotz and K. Ng* (1989)

37 Generalized Linear Models, 2nd edition *P. McCullagh and J.A. Nelder* (1989)

38 Cyclic Designs *J.A. John* (1987)

39 Analog Estimation Methods in Econometrics *C.F. Manski* (1988)

40 Subset Selection in Regression *A.J. Miller* (1990)

41 Analysis of Repeated Measures *M. Crowder and D.J. Hand* (1990)

42 Statistical Reasoning with Imprecise Probabilities *P. Walley* (1990)

43 Generalized Additive Models *T.J. Hastie and R.J. Tibshirani* (1990)

44 Inspection Errors for Attributes in Quality Control *N.L. Johnson, S. Kotz and X. Wu* (1991)

(Full details concerning this series are available from the Publishers.)

Inspection Errors for Attributes in Quality Control

NORMAN L. JOHNSON
Department of Statistics, University of North Carolina

SAMUEL KOTZ
Department of Management Science and Statistics, University of Maryland

XIZHI WU
Department of Mathematics, Nankai University

CHAPMAN & HALL
London · New York · Tokyo · Melbourne · Madras

UK	Chapman & Hall, 2-6 Boundary Row, London SE1 8HN
USA	Chapman & Hall, 29 West 35th Street, New York NY10001
JAPAN	Chapman & Hall Japan, Thomson Publishing Japan, Hirakawacho Nemoto Building, 7F, 1-7-11 Hirakawa-cho, Chiyoda-ku, Tokyo 102
AUSTRALIA	Chapman & Hall Australia, Thomas Nelson Australia, 102 Dodds Street, South Melbourne, Victoria 3205
INDIA	Chapman & Hall India, R. Seshadri, 32 Second Main Road, CIT East, Madras 600035

First edition 1991

Typeset in 10/12 Times by Thomson Press (India) Ltd, New Delhi
Printed in Great Britain by St. Edmundsbury Press,
Bury St. Edmunds, Suffolk

ISBN 0 412 38770 0

British Library Cataloguing in Publication Data

Johnson, Norman L.
Inspection Errors for Attributes in Quality Control
1. Sampling inspection
I. Title II. Kotz, Samuel III. Wu, Xizhi
658.562015195

ISBN 0 412 38770 0

Library of Congress Cataloging-in-Publication Data

Available

Contents

Preface **ix**

1 Introduction **1**

Part One: Acceptance sampling

2 Basic distributions: single sampling **11**
2.1 Introduction 11
2.2 The case $p' = 0, p < 1$ 12
2.3 The case $p' > 0, p < 1$ 15
2.4 Lots with k (> 2) strata 26

3 Basic distributions: multiple sampling **33**
3.1 Introduction 33
3.2 The case of equal error probabilities 33
3.3 The case with unequal error probabilities 36

4 Double and link sampling for acceptance **42**
4.1 Introduction 42
4.2 Double and multi-stage sampling for acceptance 42
4.3 Link sampling 48
4.4 Partial link sampling 51
4.5 Continuous sampling plans (CPS) 52

5 Multitype nonconformities **57**
5.1 Introduction 57
5.2 Inspection for one NC type among two 58
5.3 Inspection for one NC type in multitype nonconformity case 61
5.4 Multitype nonconformities, the general case 66

Part Two: Identification of nonconforming items

6 (Dorfman) group testing 77
6.1 Introduction 77
6.2 Original Dorfman procedure 80
6.3 Hierarchal screening 85
6.4 Dorfman procedures with random group size and two NC types 91

7 Dorfman–Sterrett group testing and its extensions 97
7.1 Introduction 97
7.2 Dorfman–Sterrett procedure 98
7.3 Modified Dorfman–Sterrett screening procedures 106
7.4 Randomized-sequential group testing procedures 111

8 Curtailed Dorfman-type procedures 117
8.1 Introduction and notation 117
8.2 Curtailed group testing 119
8.3 Tables and comments 125

9 Graff and Roeloffs' approach and related procedures 134
9.1 Introduction 134
9.2 Graff and Roeloffs' modification 135
9.3 Modification for hierarchal screening 137

10 Binary search 144
10.1 Introduction 144
10.2 Evaluation of $P(y, 2^k)$ 145
10.3 Evaluation of $P(y, n)$ $n \neq 2^k$ 150

Part Three: Miscellaneous

11 Estimation of error probabilities 155
11.1 Introduction 155
11.2 Estimation with p' negligible 155
11.3 General estimation by moments 158
11.4 Estimation with two inspection devices 162
11.5 Moment estimation on multitype nonconformities 169
11.6 Testing hypothesis $p = 1$ (assuming $p' = 0$) 175

12 Stratified populations: grading **181**
12.1 Grading 181
12.2 A grading problem on multitype nonconformities 183
12.3 Quota fulfilment with errors in stratification 185

Appendix A Note on computer programs **192**

Apppendix B The detection of defective members of large populations **194**

References **200**

Index **210**

Preface

Every book should have a motivation and a goal. We would like to share with our readers some reasons which have led us to allot a considerable part of our time and effort to the production of this monograph. We believe this book to originate from two 'godfathers'. The first took the form of the following, apparently straightforward, problem posed to us in 1979 by Horton L. Sorkin (at that time in the Department of Accounting, University of Maryland at College Park).

'There are N accounts, D with mistakes and $N - D$ mistake-free. The probability that an accountant detects that an account is erroneous is $p\ (<1)$. The accountant checks $n\ (<N)$ accounts, chosen at random. What is the probability that he will find Z of them to be erroneous?'

This simple question, which H. L. Sorkin had solved a few years earlier in his PhD dissertation, leads to a compound distribution which had not (to our knowledge) been discussed heretofore in the statistical literature. Our interest was aroused by this, and we carried out a further literature search, and found that, although inspection error had indeed appeared in quality control literature during the previous twenty years or so, distributional aspects had been, by and large, neglected. Sorkin's model, of course, admits of a variety of extensions and generalizations, in particular, allowing for false positive as well as false negative decisions – and these constitute the major part of the contents of the first few chapters of this monograph.

The second 'godfather' is the classical paper by R. E. Dorfman in the 1943 *Annals of Mathematical Statistics* (Vol. 14), wherein he not only initiated the now voluminous literature (greatly influenced by M. Sobel and his co-workers, e.g. 'Group testing to eliminate efficiently all defectives in a binomial sample', by M. Sobel and P. A. Groll in *Bell System Technical Journal*, Vol. 38, in 1959) on the theory of 'group testing' for nonconforming ('defective') items, but also described the successful practical application of the method by the US Army in World War II for testing blood samples from inductees for syphilis antibodies. (We reprint this classical paper in Appendix B.) Dorfman's ideas, and Sterrett's modification some fourteen years later (*Annals of Mathematical Statistics*, Vol. 28, 1957) provided further models for which we could study the effects of errors. The consequent studies constitute the subject matter of Chapters 6–9.

We believe that the possibility of savings in the amount of inspection in some of the modifications of group testing procedures is worthy of consideration in the construction of certain inspection schemes in industry, and especially in medicine. We hope that many quality assurance practitioners will find the material in this monograph of substantial potential value and interest to justify spending time on studying it. This is the main reason why we have tried to keep the mathematical level fairly low. The techniques are essentially elementary, although the necessary analysis is sometimes rather elaborate. We have often presented details of the latter, in order to justify our conclusions, but have endeavoured to indicate where detailed study is not essential to an understanding of the results. To facilitate understanding, we have included diagrammatic representations of procedures in most chapters. We have also provided extensive tables to assist appreciation of the numerical consequences of the formulas in a range of special cases. It is not practicable to cover anything like a complete range of parameter values in these tables – a book twice the size would be necessary – but *ad hoc* calculation is quite straightforward for any specific case.

A secondary motivation for this monograph is to provide a text on some modern developments in discrete applied probability theory. As just noted, the methods used in the derivation of our results, although elementary, are by no means trivial, requiring a certain amount of maturity, and familiarity with probabilistic ways of thinking. (Note especially the need to discriminate between different conditional distributions.) We feel strongly that the monograph can be of interest and usefulness to persons wishing to sharpen and broaden their technical abilities in regard to derivation and analysis of discrete distributions.

The nature of the monograph does not allow us to include formal exercises. There are, however, many numerical examples, especially in the earlier chapters. We trust that our effort in compiling this monograph will be rewarded by a heightened awareness, among practitioners and developers of quality control methodology, of a potentially useful and timely area, worthy of further study and development.

We take this opportunity to express our gratitude to those who have contributed – directly or indirectly – to the preparation of this volume.

First and foremost we thank our wives for understanding that writing yet another book can involve some sacrificial weekends and occasional burning the midnight oil during the week.

We are grateful to Dr R.D. Lamone, Dean of the College of Business and Management and Dr B.L. Golden, Chairperson of the Department of Management Science and Statistics at the University of Maryland, College Park; to Dr S. Cambanis, Chairperson of the Department of Statistics, University of North Carolina at Chapel Hill; and to Dr Hu Go-Din, Director of the Institute of Mathematics at Nankai University, Tianjin, China for their encouragement and provision of facilities enabling us to complete this project.

We offer special thanks to Ms Elizabeth Johnston, Senior Editor, and the sub-editing and production department at Chapman & Hall for their guidance and experienced professional handling of all stages of the production process.

Finally, we are happy to acknowledge the indispensable help of Dr Robert N. Rodriguez of the SAS Institute, Cary, North Carolina with regard to computational aspects of many of the results derived in this book. He has taken a keen interest in our work and has carried out computations of many tables with great efficiency and lightning speed.

Norman L. Johnson Samuel Kotz Xizhi Wu

CHAPTER 1

Introduction

Inspection, of various kinds, has been an essential component of quality control methodology for some sixty years. Originating with the pioneering work of W.A. Shewhart (1926, 1931) and Dodge and Romig (1929), inspection of samples has been systematically developed. In the course of time, many variants on simple sampling schemes have been developed, with the aim of improving efficiency and flexibility. Viewpoints on appropriate location of sampling inspections in the production and merchandising processes have changed over the years, but until recently the utility of sampling inspection has been unchallenged. Relatively recently, questions on the usefulness of some acceptance sampling procedures have arisen. It is noteworthy that in the arguments about these matters little attention has been paid to the possible effects of errors in drawing conclusions from inspection procedures. It would, however, be wildly idealistic to suppose that erroneous conclusions will never be reached; that neither instruments nor human observers will ever make mistakes (see Armstrong (1982), who elaborates on this point).

We strongly believe that discussion of sampling procedures needs to take into account the possible effects of errors in carrying out the procedures and interpreting the results obtained. This comment applies equally to procedures designed to identify each individual in a set that does not conform to specified requirements.

The earliest reference to inspection errors in quality control literature seems to be J.M. Juran's note in *Mechanical Engineering* (1935) (Vol. 57, pp. 643–644). In a pioneering paper in 1946, M. Lavin modified classical sampling inspection plans by introducing the concept of 'effective' fraction defective ($\tilde{\omega}$) related to the true fraction defective (ω) by the formula

$$\tilde{\omega} = \omega p + (1 - \omega)p'$$

where p is the probability of correctly classifying a defective item, and p' is the probability of (incorrectly) classifying a non-defective item as defective. Lavin noted that p and p' might be functions of ω, but limited his discussion to cases of constant p and p', and cases where $p = 1 - c + b\omega$ and $p' = a + bp$.

Lavin considered changes in sampling schemes needed to compensate for inspection errors. He presented results of approximate probabilities of acceptance based on constant p $(= 0{\cdot}95)$ and p' $(= 0{\cdot}02)$ for single sampling schemes with a sample of size 235 when up to seven defective items lead to acceptance (a) if no errors were present, and (b) if errors with stated p and p' are present. (Table 1.1) He then showed that with a sample size of 460 and a maximum acceptable number of defects of 23, about the same probabilities of acceptance could be achieved when subject to error, when $\omega = 0.02$, 0.05, as for the original scheme when there was no inspection error. (Table 1.2)

Lavin's broad conclusion was that it is 'largely sufficient to eliminate the rating of non-defective product as defective in order

Table 1.1 *Single sample plan, $n = 235$, $c = 7$*

No inspection errors ($p = 1$, $p' = 0$)		*$p = 0{\cdot}95$, $p' = 0{\cdot}02$*	
ω *True proportion defective*	*Approximate probability of accepting lot*	$\tilde{\omega}$ *Effective proportion defective*	*Approximate effective probability of accepting lot*
0·000	1·000	0·020	0·90
0·005	0·999	0·025	0·76
0·010	0·997	0·029	0·63
0·015	0·97	0·034	0·45
0·020	0·90	0·039	0·30
0·025	0·76	0·043	0·21
0·030	0·60	0·048	0·13
0·035	0·42	0·053	0·07
0·040	0·27	0·057	0·04
0·045	0·17	0·062	0·02
0·050	0·10	0·066	0·015
0·060	0·03	0·076	0·003
0·070	0·008	0·085	0·001
0·080	0·001	0·094	0·000

Table 1.2 *Single sample plan,* $n = 460$, $c = 23$, *with* $p = 0{\cdot}95$, $p' = 0{\cdot}02$

ω *True proportion defective*	$\tilde{\omega}$ *Effective proportion defective*	*Effective probability of accepting lot*
0·000	0·020	0·999+
0·005	0·025	0·999
0·010	0·029	0·995
0·015	0·034	0·97
0·020	0·039	0·90
0·025	0·043	0·80
0·030	0·048	0·61
0·035	0·053	0·44
0·040	0·057	0·32
0·045	0·062	0·18
0·050	0·066	0·11
0·060	0·076	0·02
0·070	0·085	0·003
0·080	0·094	0·000+

to have assurances that the guarantees of the plan are substantially obtained'.

Lavin's work was generally unnoticed in quality control literature for almost two decades. As late as the middle 1960s, Master's theses (Davis (1966), McKnight (1967)) were dealing with problems addressed in Lavin's paper.

Two other seminal papers in the field are by G. Minton (1969, 1972). Minton was concerned with the effects of inspection error (and correction error) on quality control of census data-processing activities, with special emphasis on the error of classifying defectives as non-defectives, made by inspectors called 'verifiers'. He noted that the percentage of misclassifications appeared to be a fraction of the difficulty of the job, and could range from about 2% for certain matching activities to about 70% (!) in coding sample verification.

In 1968 J.M. Juran wrote an influential note in *Quality Progress*, aiming, once again, at drawing the attention of practitioners to the importance of allowing for errors in inspection.

In the literature on medically oriented applications, the related problem of misclassification in 2 × 2 tables was considered (independently) by Bross (1954) and Rubin *et al.* (1958). These articles addressed the problem of choice between a more expensive medical examination (assumed to classify without error) and a cheaper 'interview' method. Bross's work was extended by Newell (1962) and Diamond and Lilienfeld (1962).

Extensive numerical studies of errors in measurement with binomial data were conducted by the US Bureau of the Census and published in 1964. (These were based on re-interview studies of extensive housing and population characteristics measured in the 1960 US Census.)

In the early and middle 1970s problems associated with the effects of errors in acceptance sampling on probabilities of acceptance and average outgoing quality finally received attention from practitioners and researchers. We believe that the first PhD dissertation on this topic was by R.D. Collins (1974). Papers by Case *et al.* (1973, 1975) extended earlier work of Ayoub *et al.* (1970) on the topic, taking into consideration nine different rectification inspection schemes. The same authors (Bennett *et al.* (1974)) also investigated the effects of inspection errors on cost-based quality control systems (rather than systems based on purely statistical measures). This work was extended by Hoag *et al.* (1975) to sequential sampling inspection for attributes. Dorris and Foote (1978) give a useful survey covering the period up to 1978. (Regrettably, the work of Singh (1964), and his later paper with Dutta (1976), did not receive sufficient attention.)

There have been several studies on the magnitude and possible causes of inspection error, and some discussion of ways in which inspection conditions might be controlled to reduce probabilities of error (Hayes (1950), Jacobson (1952), McKenzie (1958), Rubin *et al.* (1958), Newell (1962), Harris (1968), Drury and Addison (1973), Sinclair (1978), Rahali and Foote (1982), Hwang (1984), Smith and Duvier (1984) and more recently, Maghsoodloo (1987) and Toms and Patrick (1989)). Bauer (1987, page 7) gives an interesting table of references reporting values of $(1-p)$ up to 0·90 (!) with many cases up to 0·30 and values of p' up to 0·41 with many cases up to 10%.

In recent years, the results of Case *et al.* (1975) have been further extended to a number of sampling plans including the single sampling acceptance rectification plan S3-L3 (Ghosh (1985)) (in which apparent defectives are replaced whenever they are found) and policy S2-L2 (under which rejected lots are screened 100%, and apparent defectives, found during sampling or screening, are removed and not replaced, allowing for a 'skip-lot plan' (Carr (1982)) (under which some lots are not inspected if certain requirements are met). Carr also refuted some simple-minded, but erroneous 'remedies' (compensation for inspection error) such as those proposed by Wooding (1979).

A (Bayesian) algorithm, incorporating inspection errors, to determine optimal parameter values for Bayesian single sampling attribute plans was developed by Moskowitz and Fink (1977). In experimental assessment of the algorithm, a binomial sampling distribution was assumed and three error levels were considered: (a) no errors; (b) 'low' errors ($p=0{\cdot}895$; $p'=0{\cdot}055$); and (c) 'high' errors ($p=0{\cdot}625$; $p'=0{\cdot}156$). Their overall conclusion was that 'neglecting inspection errors in developing a sampling plan can be costly', quoting the following figures for a typical example in which the total expected costs per lot were: for (a) \$258; for (b) \$264; and for (c) \$282; using optimal sampling plans in each case, which were as follows

	(a)	(b)	(c)
Sample size	68	60	53
Maximum acceptable defectives	7	9	11

Collins and Case (1976) did consider finite lot sizes, but only to show that if the lots themselves were chosen at random from production with constant proportion (ω) of nonconforming (NC) items, then the overall distribution of the number of NC decisions would be binomial with parameters n (sample size) and $\tilde{\omega}$ (the parameter introduced by Lavin (1946)), as in the case of 'infinite' lot size. Generally, these earlier studies did not take the finiteness of lot size into account.

As we have remarked in the Preface, our interest in these topics was stimulated by a query from Horton L. Sorkin, who was studying the effects of errors in accounting procedures. The results of his enquiries were summarized in his PhD thesis, and published in Sorkin (1978). During the past decade, two of us (Johnson and Kotz) have published (sometimes jointly with other authors) about twenty papers on the effects of errors on various sampling procedures, with special emphasis on group screening. A large part of this book represents a consolidation, and in some cases, refinement, of the results obtained in these papers. Efforts have been made to include work of other authors in this field, but we must apologize for possible inadvertent omissions (and, perhaps, misrepresentations). Indeed, in a book on these topics, it is especially appropriate to keep in mind the possibility of authors' errors!

We now describe the contents of this book in outline.

As we have noted, this book is concerned with the effects of errors in observations on results of certain statistical procedures. It is limited to situations wherein attributes are measured, rather than characters such length, weight, pressure etc., which are traditionally represented (as a convenient approximation) by continuous variables. By contrast, the results of an observation on an attribute – its presence or absence – can be represented by a binary variable taking values 1 or 0 only.

We will eventually consider situations in which several attributes are measured on each individual, but initially we will discuss problems in which there is only a single attribute, which will be conventionally described as 'nonconformity'. For convenience and brevity we will use the term 'item' for an individual. A **nonconforming** (**conforming**) individual will be described as an 'NC (C) item'.

In the simple situation just described, there are only two possible errors in observation.

1. An NC item may be classified as C.
2. A C item may be classified as NC.

The symbols $(1-p)$, p' will be used for the respective probabilities of these two kinds of error. This can be summarized formally as

$$\Pr[\mathrm{C}|\mathrm{NC}] = 1 - p \text{ (i.e. } \Pr[\mathrm{NC}|\mathrm{NC}] = p); \quad \Pr[\mathrm{NC}|\mathrm{C}] = p'.$$

The symbols p, p' each denote probabilities of classification as NC. In specific cases, identifying symbols – e.g. subscripts – will be attached, as needed, to p and p' to indicate to which parts of a procedure they apply. For example, in multi-stage procedures, a subscript might identify the stage of the procedure to which the symbol refers.

The symbol ω will be used to denote the proportion of items in a population (e.g. a lot) which are NC. If more than one attribute is to be measured, appropriate subscripts will be used to identify particular attributes.

Quite small values of p' may lead to markedly untoward effects. The percentage of C items among all items declared NC is

$$100(1-\omega)p'/\{\omega p + (1-\omega)p'\}\%.$$

If ω, the proportion of NC items in the population, is 0·01 and p, the probability of detecting an NC item, is 0·99, then a value of 50% would be produced by p' equal to 0·01. If p' were equal to 0·03, the percentage of C items among those declared NC would be 75%.

Although our main attention will be concentrated on the consequences of errors in observation, there is also some discussion of problems of estimation, in particular estimation of the error probabilities, $(1 - p)$ and p', and also of design problems arising in efforts to reduce the possible effects of errors.

Most of the procedures discussed fall into one of two main classes:

1. acceptance sampling procedures ('sampling plans');
2. search for NC items.

These form the contents of Parts One and Two. In Part Three several procedures, not falling into either (1) or (2), are the subjects of discussion.

The essential distinction between (1) and (2) lies in the form of decision to be reached. In (1) decisions relate to a population as a whole (e.g. all items in a lot) and are of the nature of overall approval ('acceptance') or disapproval ('rejection'). In (2) decisions are individual; each item in the population is to be classified as NC or C.

Indices of the properties of procedures should measure the probabilities of error in final decisions and the expected number of observations to be needed, and these will be what is primarily determined in this book. There is some attention to actual costs, indicating where this is likely to be especially necessary. Such situations occur when unit sampling costs may vary according to the stage in a procedure where they arise, and when it is important to consider the expected costs arising from incorrect decisions. However, attention to questions of cost is generally limited. Bauer (1987, Chapter 4) studies consequences of a cost structure devised by Collins (1974) in relation to optimal (minimum expected cost) choice of sample size and acceptance number for single sampling procedures (see Chapter 3). (Bauer (1987) reports that numerical results he obtained in special cases considered by Collins (1974) are at variance with those obtained by the latter.)

It is important to keep in mind the fact that the applicability of the various procedures in this book depends on the nature of the material. For example, if observation results in destruction of the item observed (e.g. in breakage strength tests) it is not possible to take repeated observations. Also it may not be possible to test for the presence of at least one NC item in a group by a single observation on the group as a whole.

Most of the discussions apply to situations in which ω, p and p' do not vary with time. Some attention will, however, be paid to problems in which the last two parameters do vary. Another general assumption is that results of observations from different items or on the same item at different times are mutually independent.

In a few cases, formulas for indices of properties of procedures are sufficiently simple ('transparent') for their practical effects to be gauged without extensive numerical calculation. Mostly, however, formulas are rather elaborate (although involving only quite simple functions). Tables are provided giving values in some situations but it is not practicable to include more than a representative selection from the wide range of possibilities. It will often be necessary to make calculations custom-tailored to each specific problem. To assist in such calculations, computer programs are available for the more important cases. Some of these on listed in Appendix A.

The book by Bauer (1987) provides, in many ways, a useful complement to the present work. Apart from discussion of influence of costs (Chapter 4), it devotes Chapter 3 to the construction of 'error-compensating' single sampling inspection plans on similar lines to Moskowitz and Fink (1977). There are extensive tables of sample size (n) and maximum number (c) of items declared NC for acceptance, under inspection errors which will provide probabilities of acceptance and rejection equivalent to these obtained with $n = 18$ and $c = 4$ when inspection is error-free. (Gunther (1969), Guthrie and Johns (1959) and Hailey (1980) are also relevant to this work.)

PART ONE

Acceptance sampling

CHAPTER 2

Basic distributions: single sampling

2.1 Introduction

In sections 2.2 and 2.3, we will assume that there is a population (lot) of size N, containing D nonconforming (NC) items. We will confine ourselves to the single sample case, i.e. a random sample of size n will be taken *without* replacement, and all items in the sample are inspected. It is assumed that all decisions following the inspection are mutually independent from item to item. Afterwards, each inspected item is declared to be NC or C. The inspection might very well not be perfect. We will consider two kinds of errors, as mentioned in Chapter 1. Let p denote the probability that an NC item is (correctly) declared NC, and p' the probability that a C item is (incorrectly) identified as NC. We also denote the actual (unknown) number of NC items in the sample by Y, a random variable, and let Z denote the number of items in the sample declared to be NC (correctly or incorrectly) as the result of inspection.

When the size of the population is large, i.e. N is effectively infinite ($N \to \infty$), we will also assume that $D \to \infty$ and $D/N \to \omega$, say. For a finite lot size this would give the same result as sampling *with* replacement.

The distribution of Z is our present interest. It will be obtained by compounding the conditional distribution of Z, given Y, by the distribution of Y. We will also obtain moments of Z, using descending factorial moments.

In section 2.4, we will consider a more general case, where, instead of the simple division which splits the lot into D NC and $(N-D)$ C items, respectively, we have a lot with k strata or classes. Our inspection is to identify which items are NC. We assume that the probability that an item from the jth stratum is identified as NC is p_j for $j = 1, \ldots, k$. We will still consider single sampling.

The cases discussed in this chapter are very basic, and one can easily make generalizations following the same techniques. The symbols used here will be valid through most of the book, although some modifications for special situations are introduced from time to time.

2.2 The case $p' = 0$, $p < 1$

Distributions

Before the general case, we will first consider the case $p' = 0$. This means that all items declared NC must indeed by NC. The probability that there are $Y = y$ NC items in a sample of size n is

$$\Pr[Y = y] = \frac{\binom{n}{y}\binom{N-n}{D-y}}{\binom{N}{D}} = \frac{\binom{D}{y}\binom{N-D}{n-y}}{\binom{N}{n}},$$

$$\max(0, n - N + D) \leqslant y \leqslant \min(n, D). \tag{2.1a}$$

Symbolically,

$$Y \sim \text{Hypg}(n; D, N), \tag{2.1b}$$

where Hypg denotes the hypergeometric distribution. The conditional probability of 'detecting' z NC items among these y truly NC items is binomial with parameters (y, p):

$$\Pr[Z = z \mid y] = \binom{y}{z} p^z (1-p)^{y-z}, \qquad (z = 0, 1, \ldots, y). \tag{2.2a}$$

Symbolically,

$$Z \mid y \sim \text{Bin}(y, p). \tag{2.2b}$$

Therefore, the overall distribution of Z is

$$\Pr[Z = z] = \left\{ \sum_{y \leqslant z} \binom{D}{y}\binom{N-D}{n-y}\binom{y}{z} p^z (1-p)^{y-z} \right\} \Big/ \binom{N}{n},$$

$$\max(0, n - N + D) \leqslant z \leqslant \min(n, D), \tag{2.3a}$$

or, symbolically,

$$Z \sim \text{Bin}(Y, p) \bigwedge_{Y} \text{Hypg}(n; D, N), \tag{2.3b}$$

where $\sim$ means 'is distributed as'. The symbol $\wedge$ is used routinely to denote *mixing* or *compounding* with respect to the variable(s) below it, with the distribution shown to the right of it.

Moments

Factorial moments are convenient for calculating moments (in particular, expectations and variances) especially for binomial, Poisson and hypergeometric distributions and mixtures thereof. The rth **(descending) factorial** of α is

$$\alpha^{(r)} \equiv \alpha(\alpha - 1)\cdots(\alpha - r + 1),$$

and the rth **(descending) factorial moment** of a random variable X is

$$\mu_{(r)}(X) \equiv E[X^{(r)}],$$

the expected value of $X^{(r)}$. It is well known that, if X has a Bin(m, θ) distribution,

$$\mu_{(r)}(X) = m^{(r)}\theta^r. \tag{2.4a}$$

If X has a Hypg$(m; \delta, v)$ distribution,

$$\mu_{(r)}(X) = m^{(r)}\delta^{(r)}/v^{(r)}, \tag{2.4b}$$

and

$$E[X^{(r)}(m - X)^{(s)}] = m^{(r+s)}\delta^{(r)}(v - \delta)^{(s)}/v^{(r+s)}. \tag{2.4c}$$

For our Z, conditioned on $Y = y$, we thus have

$$E[Z^{(s)} \mid Y = y] = p^s y^{(s)}, \tag{2.5}$$

and

$$\mu_{(s)}(Z) = p^s E[Y^{(s)}] = p^s n^{(s)} D^{(s)}/N^{(s)} \qquad \text{(by (2.4b))}. \tag{2.6}$$

In particular, for $s = 1$, we have the expected value

$$E[Z] = E[Z^{(1)}] = pn\frac{D}{N} = n\frac{pD}{N} = n\bar{p}, \tag{2.7}$$

where $\bar{p} = pD/N$ is the probability that a *randomly chosen* item will be classified as NC. We also have, from the factorial moments,

$$\begin{aligned} \operatorname{var}(Z) &= p^2 n\frac{(N - n)D}{N - 1}\left(1 - \frac{D}{N}\right) + p(1 - p)\frac{nD}{N} \\ &= p^2 \operatorname{var}(Z \mid p = 1) + p(1 - p)\frac{nD}{N}, \end{aligned} \tag{2.8a}$$

using $\text{var}(Z) = \mu_{(2)}(Z) + E[Z] - E[Z]^2$. (Note that $\text{var}(Z|p=1)$ is the variance of a hypergeometric distribution with parameters $(n; D, N)$.) We also have two alternative forms for $\text{var}(Z)$:

$$\text{var}(Z) = \frac{n(N-n)}{N-1}\bar{p}(1-\bar{p}) + \frac{n-1}{N-1}p(1-p)\frac{nD}{N} \tag{2.8b}$$

or

$$\text{var}(Z) = n\bar{p}(1-\bar{p}) - \frac{n(n-1)}{N(N-1)}p^2 D\left(1 - \frac{D}{N}\right). \tag{2.8c}$$

Expressions (2.8b) and (2.8c) indicate that $\text{var}(Z)$ is not less than the variance of the hypergeometric with parameters (n, pD, N), and cannot be greater than that of the binomial with parameters $(n, \bar{p})$. (These two variances are

$$\frac{n(N-n)}{N-1}\bar{p}(1-\bar{p}) \qquad \text{and} \qquad n\bar{p}(1-\bar{p}),$$

respectively.)

Waiting time

Suppose that items are taken one at a time, from a lot containing N items (without replacement) and inspected one after the other. The process is continued until m items have been declared NC. What is the distribution of the number, G, of items inspected to obtain these m NC decisions?

The answer is readily derived from the distribution (2.3a) of Z. We have

$$\Pr[G > g] = \Pr[Z < m] \tag{2.9}$$

where Z has the distribution (2.3a) with n replaced by g. It is, of course, possible that all N items in the lot might be inspected without obtaining m NC decisions, even though $D \geqslant m$. In fact

$$\Pr[G \leqslant N] = \Pr[Z \geqslant m]$$

with Z having a binomial (D, p) distribution. The distribution of G is therefore improper.

If only a sample of n items is inspected, the resultant distribution of G is still obtainable from (2.9), but it will be truncated to values

$G \leqslant n$. The distribution will, *a fortiori*, be improper, since

$$\Pr[G \leqslant n] = \Pr(Z \geqslant m] < 1$$

with Z having distribution (2.3a).

A fuller discussion of the distribution of G, in more general cases, is presented in section 2.3.

2.3 The case $p' > 0, p < 1$

Distributions

As in the case $p' = 0$, $p = 1$, discussed in the preceding section the distribution of Y, the true number of NC items in the sample of size n, is hypergeometric with parameters $(n; D, N)$. (See (2.1a) and (2.1b).) Furthermore, if $N \to \infty$ and $D \to \infty$ with $D/N \to \omega$ (or if sampling is with replacement, and $D/N = \omega$), Y has a binomial distribution with parameters (n, ω):

$$\Pr(Y = y|n) = \binom{n}{y}\omega^y(1 - \omega)^{n-y} \qquad (0 \leqslant y \leqslant n). \qquad (2.10)$$

The total number, Z, of items classified (correctly or incorrectly) as NC is the sum of two variables:

$$Z = W + W', \qquad (2.11)$$

where W is the number of NC items classified as NC, and W' the number of C items classified as NC. W and W' are mutually independent and each is binomially distributed, given Y, i.e.

$$W|Y \sim \text{Bin}(Y, p),$$

$$W'|Y \sim \text{Bin}(n - Y, p').$$

The conditional distribution of Z, given Y, is

$$Z|Y \sim \text{Bin}(Y, p) * \text{Bin}(n - Y, p') \qquad (2.12a)$$

where $*$ stands for **convolution**. Explicitly,

$$\Pr[Z = z|Y] \equiv \Pr[Z = z|Y; n; D, N]$$

$$= \sum_w \binom{Y}{w} p^w(1 - p)^{Y-w}\binom{n - Y}{z - w} p'^{z-w}(1 - p')^{n-Y-z+w} \qquad (2.12b)$$

where $\max(0, z - n + Y) \leqslant w \leqslant \min(Y, z)$.

For finite N, the overall distribution of Z can then be obtained by compounding the conditional distribution of Z, given Y, by the distribution of Y. We have

$$Z \sim [\text{Bin}(Y, p) * \text{Bin}(n - Y, p')] \bigwedge_{Y} \text{Hypg}(n; D, N). \qquad (2.13\text{a})$$

Explicitly,

$$\begin{aligned}\Pr[Z = z] &\equiv \Pr[Z = z \mid n; D, N; p, p'] \\ &= \binom{N}{n}^{-1} \sum_{y} \binom{D}{y}\binom{N-D}{n-y} \sum_{w} \binom{y}{w} p^{w}(1-p)^{y-w} \\ &\quad \times \binom{n-y}{z-w} p'^{z-w}(1-p')^{n-y-z+w}, \qquad (2.13\text{b})\end{aligned}$$

where limits for y and w are given by

$$\max(0, n - N + D) \leqslant y \leqslant \min(n, D) \quad \text{and} \quad \max(0, z - n + y) \leqslant w \leqslant \min(y, z).$$

As $N \to \infty$ (cf. (2.10)), we have

$$\begin{aligned}Z &\sim [\text{Bin}(Y, p) * \text{Bin}(n - Y, p')] \bigwedge_{Y} \text{Bin}(n, \omega) \\ &= \text{Bin}(n, \bar{p}), \qquad (2.14)\end{aligned}$$

where $\bar{p} = \omega p + (1 - \omega)p'$ is the probability of classifying a *randomly chosen* item as NC. For finite N, with DN^{-1} replacing ω, and

$$\bar{p} = DN^{-1}p + (1 - DN^{-1})p',$$

distribution (2.14) gives an approximation for (2.13a).

Moments

From the algebraic identity

$$Z^{(r)} = (W + W')^{(r)} = \sum_{j=0}^{r} \binom{r}{j} W^{(j)} W'^{(r-j)},$$

one can get the rth descending factorial moment of Z conditioned on Y:

$$\mu_{(r)}(Z \mid Y) = \sum_{j=0}^{r} \binom{r}{j} p^{j} p'^{r-j} Y^{(j)} (n - Y)^{(r-j)}. \qquad (2.15)$$

Hence

$$\mu_{(r)}(Z) = \frac{n^{(r)}}{N^{(r)}} \sum_{j=0}^{r} \binom{r}{j} p^j p'^{r-j} D^{(j)}(N-D)^{(r-j)}. \tag{2.16}$$

As a consequence of (2.16), (cf. (2.7) and (2.8a–c)), we obtain

$$E[Z] = n\frac{Dp + (N-D)p'}{N} = n\{\omega p + (1-\omega)p'\} = n\bar{p} \qquad \text{(cf. (2.7))} \tag{2.17}$$

and

$$\begin{aligned}\operatorname{var}(Z) &= n\bar{p}(1-\bar{p}) - \frac{n(n-1)}{(N-1)}\frac{D}{N}\left(1-\frac{D}{N}\right)(p-p')^2 \\ &= n\bar{p}(1-\bar{p}) - \frac{n(n-1)}{(N-1)}(\overline{p^2} - \bar{p}^2) \qquad \text{(cf. (2.8c))}\end{aligned} \tag{2.18}$$

where

$$\omega = D/N \qquad \text{and} \qquad \overline{p^2} = \frac{Dp^2 + (N-D)p'^2}{N}.$$

Tables

For $N = 100$, 200, ∞; $D/N = 0{\cdot}05$, $0{\cdot}1$, $0{\cdot}2$; $n = 5$, 10, 15; $p = 0{\cdot}75$ $(0{\cdot}05)$ 1.00; $p' = 0{\cdot}00$ $(0{\cdot}025)$ $0{\cdot}10$, Tables 1 and 2 of Johnson *et al.* (1985a) give values of $\Pr[Z = z]$ to four decimal places for both finite and infinite lot sizes, respectively. In the case of effectively infinite sample sizes, we replace D/N by ω. We reproduce here values for $N = 100, 200, \infty$; $D/N(\omega) = 0{\cdot}05, 0{\cdot}2$; $n = 5, 10$; $p = 0{\cdot}75, 0{\cdot}90, 1{\cdot}00$; $p' = 0{\cdot}00$, $0{\cdot}05$, $0{\cdot}10$ (Table 2.1). Results for perfect inspection correspond to $p = 1$ and $p' = 0$, and we can see the effects of imperfection from the tables.

These tables, despite the limited variety of values of N (100, 200 and ∞), give a clear picture of the effect of lot size on the distribution of Z. For example, taking D/N (or ω) $= 0{\cdot}05$, $n = 5$, $p = 0{\cdot}9$, $p' = 0{\cdot}075$, we have the following values of $\Pr[Z = z]$:

N	z = 0	1	2	3	4	5
100	0·5368	0·3583	0·0926	0·0116	0·0007	0·0000
200	0·5379	0·3564	0·0930	0·0119	0·0008	0·0000
∞	0·5391	0·3546	0·0933	0·0123	0·0008	0·0000

Table 2.1 *Distribution of Z (number of items declared nonconforming): Values of* $\Pr[Z = z] \times 10\,000$

	Lot size (N) = 100																	
Proportion nonconforming (D/N)	0.05									0.20								
$p =$	0·75			0·90			1·00			0·75			0·90			1·00		
$p' =$	0·00	0·05	0·10	0·00	0·05	0·10	0·00	0·05	0·10	0·00	0·05	0·10	0·00	0·05	0·10	0·00	0·05	0·10
Sample size (n) = 5																		
z																		
0	8236	6395	4900	7909	6129	4684	7696	5955	4544	4381	3444	2675	3634	2831	2177	3193	2471	1885
1	1655	3011	3773	1936	3182	3859	2114	3289	3912	3998	4145	4078	4168	4137	3939	4201	4072	3804
2	0106	0544	1144	0150	0628	1244	0184	0685	1312	1379	1928	2431	1798	2322	2770	2073	2567	2969
3	0003	0047	0171	0004	0059	0196	0006	0068	0215	0224	0433	0709	0364	0625	0947	0478	0773	1121
4	0001	0002	0013	–	0003	0015	–	0003	0017	0017	0047	0101	0034	0081	0157	0051	0111	0205
5	–	–	–	–	–	–	–	–	0001	–	0002	0006	0001	0004	0010	0002	0006	0015
Sample size (n) = 10																		
z																		
0	6731	4060	2384	6184	3714	2170	5838	3495	2035	1856	1149	0694	1253	0761	0450	0951	0569	0332
1	2818	3891	3711	3183	3953	3634	3394	3979	3516	3540	2860	2173	3044	2331	1695	2679	1988	1407
2	0422	1608	2557	0584	1793	2675	0702	1905	2748	2870	3093	2992	3128	3081	2786	3182	2982	2592
3	0028	0379	1028	0047	0458	1141	0064	0515	1219	1299	1915	2386	1787	2314	2635	2092	2528	2734
4	0001	0056	0267	0002	0073	0313	0003	0086	0346	0362	0751	1221	0626	1093	1588	0841	1342	1829
5	–	0006	0047	–	0008	0058	–	0009	0066	0065	0195	0419	0140	0339	0638	0215	0466	0811
6	–	–	0006	–	0001	0007	–	0001	0009	0008	0034	0098	0020	0070	0173	0035	0107	0242
7	–	–	–	–	–	0001	–	–	0001	0001	0004	0015	0002	0010	0031	0004	0016	0048
8	–	–	–	–	–	–	–	–	–	–	–	0002	–	0001	0004	–	0002	0006
9	–	–	–	–	–	–	–	–	–	–	–	–	–	–	–	–	–	–
10	–	–	–	–	–	–	–	–	–	–	–	–	–	–	–	–	–	–

Lot size $(N) = 200$

Proportion nonconforming $(D/N) = 0.05$										0.20								
$p =$	0·75			0·90			1·00			0·75			0·90			1·00		
$p' =$	0·00	0·05	0·10	0·00	0·05	0·10	0·00	0·05	0·10	0·00	0·05	0·10	0·00	0·05	0·10	0·00	0·05	0·10
Sample size $(n) = 5$																		
z																		
0	8248	6405	4906	7927	6142	4694	7717	5971	4557	4409	3466	2691	3671	2859	2198	3235	2503	1910
1	1632	2995	3762	1903	3159	3844	2075	3261	3893	3956	4117	4060	4118	4104	3918	4148	4037	3783
2	0116	0549	1145	0164	0634	1246	0200	0692	1313	1381	1923	2423	1792	2309	2755	2061	2547	2948
3	0004	0049	0173	0006	0062	0200	0008	0071	0219	0234	0442	0715	0378	0637	0955	0496	0786	1130
4	–	0002	0013	–	0003	0016	–	0004	0018	0019	0050	0104	0039	0086	0163	0058	0119	0213
5	–	–	–	–	–	–	–	–	0001	0001	0002	0006	0002	0005	0011	0003	0007	0016
Sample size $(n) = 10$																		
z																		
0	6778	4087	2399	6249	3752	2192	5915	3541	2062	1913	1183	0714	1314	0798	0471	1013	0606	0353
1	2736	3855	3698	3074	3905	3618	3268	3923	3557	3507	2856	2181	3031	2341	1713	2683	2010	1432
2	0446	1603	2545	0610	1783	2657	0727	1899	2725	2813	3051	2966	3052	3031	2760	3098	2932	2569
3	0038	0388	1030	0063	0470	1144	0085	0530	1221	1299	1899	2364	1766	2279	2598	2052	2477	2687
4	0002	0060	0272	0004	0080	0320	0006	0094	0355	0382	0762	1223	0649	1101	1582	0863	1343	1814
5	–	0006	0049	–	0009	0061	–	0011	0070	0075	0206	0429	0159	0358	0652	0240	0488	0826
6	–	–	0006	–	0001	0008	–	0001	0009	0010	0038	0103	0026	0079	0184	0045	0120	0257
7	–	–	0001	–	–	0001	–	–	0001	0001	0005	0017	0003	0012	0035	0006	0020	0054
8	–	–	–	–	–	–	–	–	–	–	–	0002	–	0001	0004	–	0002	0007
9	–	–	–	–	–	–	–	–	–	–	–	–	–	–	–	–	–	0001
10	–	–	–	–	–	–	–	–	–	–	–	–	–	–	–	–	–	—

Table 2.1 (*Contd.*)

	Lot size $(N) = \infty$																	
Proportion nonconforming $(\omega) =$	0.05									0.20								
$p =$	0·75			0·90			1·00			0·75			0·90			1·00		
$p' =$	0·00	0·05	0·10	0·00	0·05	0·10	0·00	0·05	0·10	0·00	0·05	0·10	0·00	0·05	0·10	0·00	0·05	0·10
Sample size $(n) = 5$																		
z																		
0	8260	6414	4913	7944	6155	4704	7738	5987	4569	4437	3487	2707	3707	2887	2219	3277	2536	1935
1	1609	2979	3752	1872	3137	3829	2036	3234	3874	3915	4089	4043	4069	4072	3898	4096	4003	3762
2	0125	0553	1146	0176	0639	1247	0214	0699	1314	1382	1919	2415	1786	2297	2739	2048	2529	2926
3	0005	0051	0175	0008	0065	0203	0011	0075	0223	0244	0450	0721	0392	0648	0692	0512	0798	1138
4	–	0002	0013	–	0003	0017	–	0004	0019	0022	0053	0108	0043	0091	0169	0064	0126	0221
5	–	–	–	–	–	0001	–	–	0001	0001	0002	0006	0002	0005	0012	0003	0008	0017
Sample size $(n) = 10$																		
z																		
0	6824	4113	2414	6310	3788	2213	5987	3585	2088	1969	1216	0733	1374	0834	0492	1074	0643	0374
1	2659	3821	3687	2973	3862	3603	3151	3873	3540	3474	2852	2188	3017	2351	1730	2684	2030	1456
2	0466	1597	2534	0630	1771	2639	0746	1883	2702	2759	3010	2942	2980	2984	2735	3020	2885	2548
3	0048	0396	1032	0079	0481	1146	0105	0542	1222	1298	1883	2343	1745	2244	2563	2013	2429	2642
4	0003	0064	0276	0007	0086	0326	0010	0103	0363	0401	0773	1225	0670	1108	1576	0881	1343	1798
5	–	0007	0051	–	0011	0064	0001	0013	0074	0085	0218	0439	0177	0375	0664	0264	0509	0839
6	–	0001	0006	–	0001	0009	–	0001	0010	0012	0043	0109	0032	0088	0195	0055	0134	0272
7	–	–	0001	–	–	0001	–	–	0001	0001	0006	0019	0004	0014	0039	0008	0024	0060
8	–	–	–	–	–	–	–	–	–	–	0001	0002	–	0002	0005	0001	0003	0009
9	–	–	–	–	–	–	–	–	–	–	–	–	–	–	–	–	–	0001
10	–	–	–	–	–	–	–	–	–	–	–	–	–	–	–	–	–	–

Good approximations for N greater than 100 can be obtained by harmonic interpolation with respect to N, i.e. linear interpolation with respect to N^{-1}, while corresponding extrapolation can work reasonably well with N as low as 50 or even lower. For example, taking $N = 40$, harmonic extrapolation gives

$$\Pr[Z = 0] = 0{\cdot}5368 + 3(0{\cdot}5368 - 0{\cdot}5379) = 0{\cdot}5335$$

compared with the actual value 0·5330, calculated from (2.13b).

Figures 1a–1d and 2a–2d from Johnson *et al.* (1985a) illustrate changes in distributions with changes in parameter values. The parameter values used are $n = 5$; $p = 0{\cdot}75$, $1{\cdot}00$; $p' = 0{\cdot}000$ $(0{\cdot}025)$ $0{\cdot}100$; and $D/N = \omega = 0{\cdot}05$, $0{\cdot}20$ for $N = 100$, and ∞, respectively. In each figure, the effect of changing values of p' can easily be determined, and comparison of Figures 1j and 2j for each j = a, b, c, d indicates that the increase in N from 100 to ∞ has relatively small effect. Figures 1a–1d and 2a–2d are reproduced here, as Figs. 2.1 and 2.2.

Example

A random sample of size $n = 15$ is taken without replacement from a lot of size $N = 100$. We accept the lot if Z, the number of items declared NC, is no more than 1. The probabilities, $P(A)$ say, of acceptance for different values of D, p and p' are of interest. We have from the tables:

case (a): $p = 1$ and $p' = 0$.

$$\begin{aligned}\text{For } D = 5:&\quad P(A) = 0{\cdot}4357 + 0{\cdot}4034 = 0{\cdot}8391;\\ \text{for } D = 20:&\quad P(A) = 0{\cdot}0262 + 0{\cdot}1191 = 0{\cdot}1453.\end{aligned}$$

case (b): $p = 0{\cdot}8$ and $p' = 0$.

$$\begin{aligned}\text{For } D = 5:&\quad P(A) = 0{\cdot}5221 + 0{\cdot}3689 = 0{\cdot}8910;\\ \text{for } D = 20:&\quad P(A) = 0{\cdot}0620 + 0{\cdot}2027 = 0{\cdot}2647.\end{aligned}$$

case (c): $p = 0{\cdot}8$ and $p' = 0{\cdot}025$.

$$\begin{aligned}\text{For } D = 5:&\quad P(A) = 0{\cdot}3587 + 0{\cdot}3960 = 0{\cdot}7547;\\ \text{for } D = 20:&\quad P(A) = 0{\cdot}0433 + 0{\cdot}1606 = 0{\cdot}2039.\end{aligned}$$

Comparing cases (a) and (c), one may notice that, for $D = 5$, $0{\cdot}7547 < 0{\cdot}8391$, i.e. the effect of false detection outweighs the effect of failure to detect, while for $D = 20$, we have the opposite result

(a) Pr [Z = z] for $p = 0.75$, $N = 100$, $D = 5$, $n = 5$

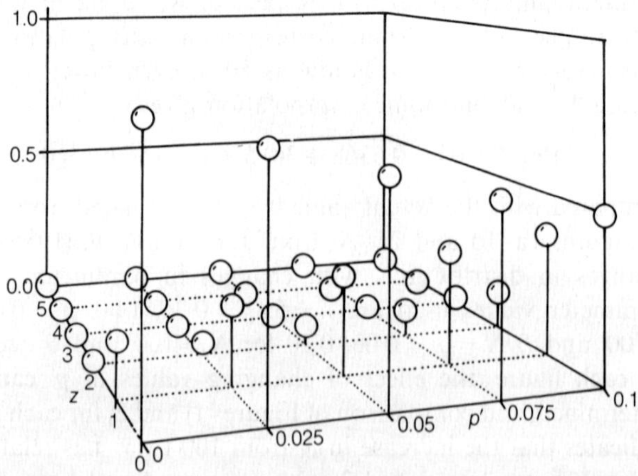

(b) Pr [Z = z] for $p = 1.00$, $N = 100$, $D = 5$, $n = 5$

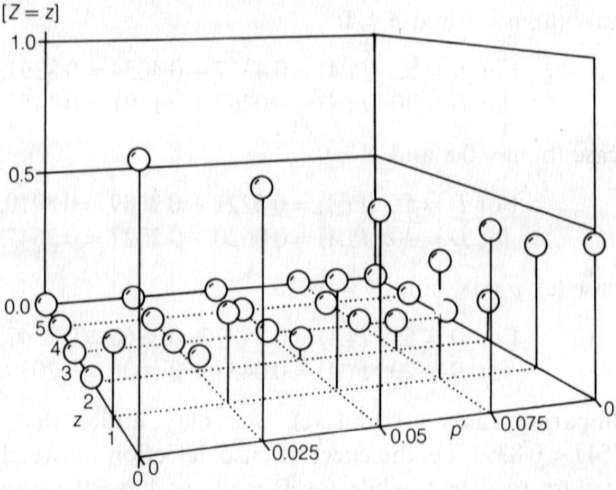

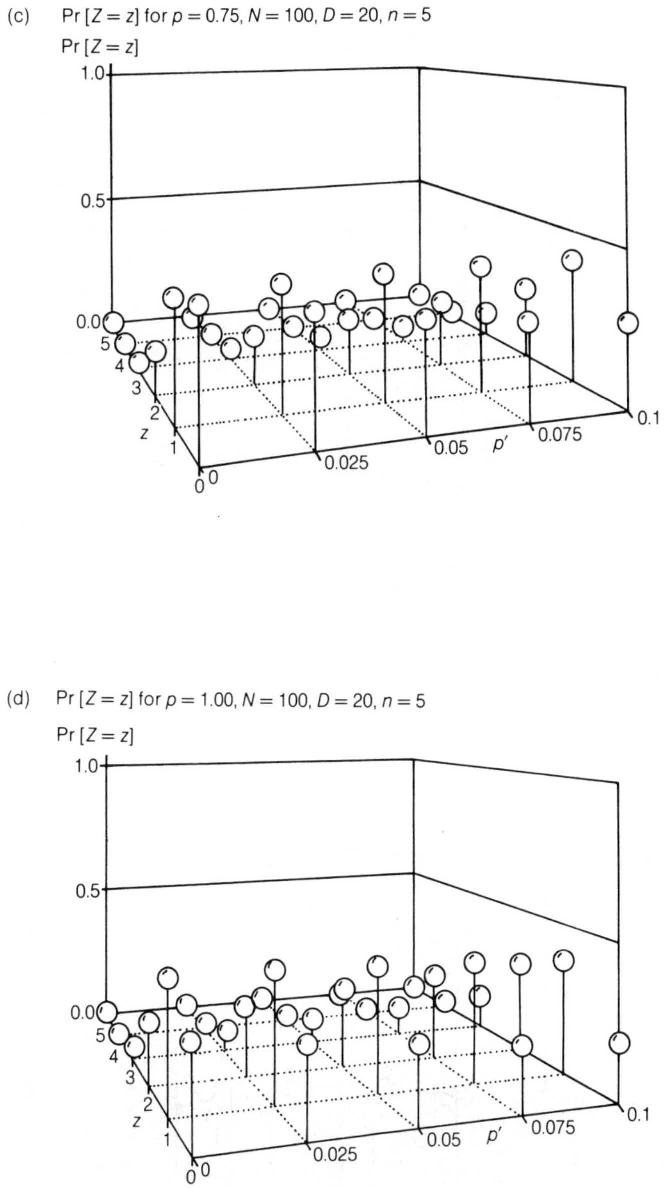

Figure 2.1 *The distribution of the number of items z classified as nonconforming when there is imperfect inspection – the finite lot size case*

(a)

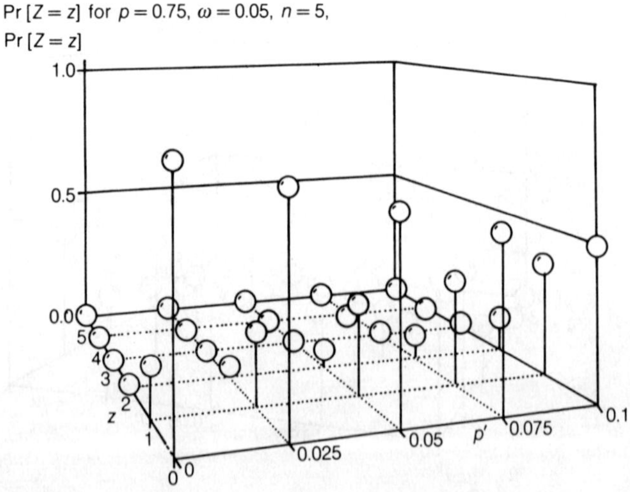
Pr [Z = z] for p = 0.75, ω = 0.05, n = 5,
Pr [Z = z]
1.0
0.5
0.0
z
5
4
3
2
1
0
0
0.025
0.05
p′
0.075
0.1

(b)

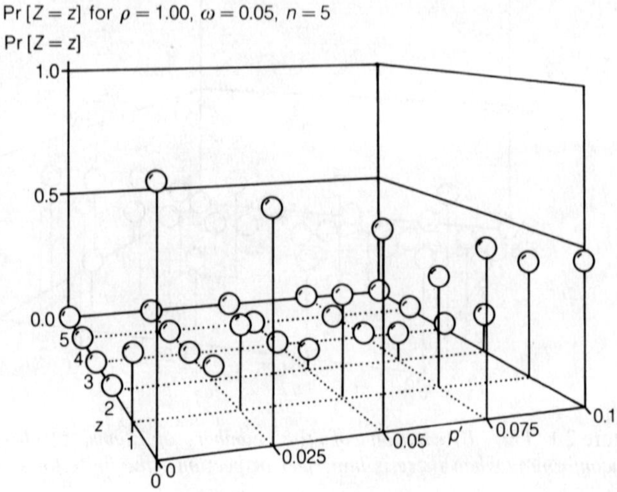
Pr [Z = z] for ρ = 1.00, ω = 0.05, n = 5
Pr [Z = z]
1.0
0.5
0.0
z
5
4
3
2
1
0
0
0.025
0.05
p′
0.075
0.1

(c) Pr [Z = z] for p = 0.75, ω = 0.20, n = 5

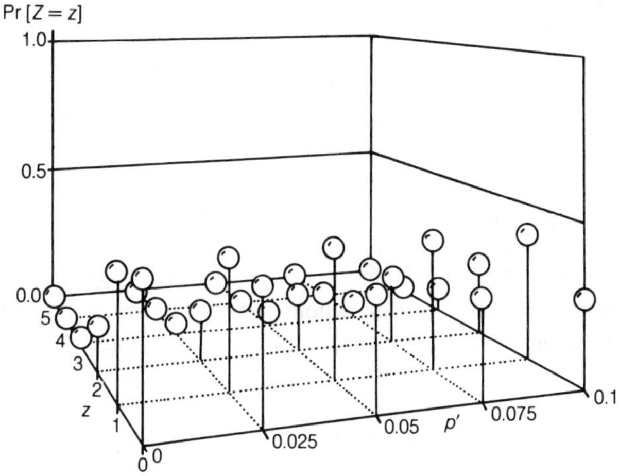

(d) Pr [Z = z] for p = 1.00, ω = 0.20, n = 5

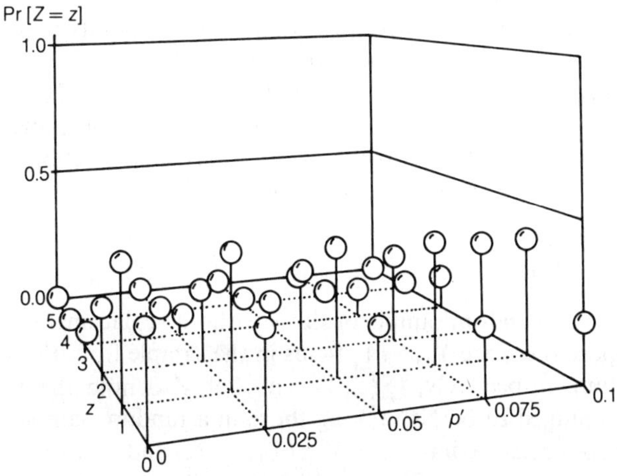

Figure 2.2 *The distribution of the number of items z classified as nonconforming when there is imperfect inspection – the finite lot size case*

since $0{\cdot}2039 > 0{\cdot}1453$. An understanding of this phenomenon can be gleaned by comparing the values of $\bar{p}$ (see (2.14)), as shown in the following table:

	$p=1$ $p'=0$	$p=0{\cdot}80$ $p'=0{\cdot}00$	$p=0{\cdot}80$ $p'=0{\cdot}025$
For $D=5$	0·05	0·04	0·06275
For $D=20$	0·20	0·16	0·18

Note that $0{\cdot}06275 > 0{\cdot}05$, but $0{\cdot}18 < 0{\cdot}20$. A simpler, intuitive explanation is to note that when there are more NC items, failure to detect them becomes more important.

2.4 Lots with k (> 2) strata

We now consider a more general situation, with a lot of N items, composed of k subsets, which we call **classes** or **strata,**

$$\Pi_1, \Pi_2, \ldots, \Pi_k$$

say. Each stratum Π_i contains N_i items ($\sum_{i=1}^{k} N_i = N$); and we suppose that the probability that an item from the ith stratum is judged to be NC is p_i.

The situation discussed in section 2.3 corresponds to $k=2$, with all items in stratum Π_1 being truly NC and all in stratum Π_2 truly C. Then $p_1 = p$ and $p_2 = p'$. In applications with $k>2$, the items in different strata might possess different degrees or expressions of nonconformity. See Chapter 12 – in particular sections 12.1 and 12.2 – for a detailed analysis of this model.

Distributions

We take a random sample of size n without replacement from a lot. Suppose there are $Y_1, \ldots, Y_k$ items in the sample from the 1st,…, kth stratum, respectively, ($\sum_{i=1}^{k} Y_i = n$). Let Z denote the number of items judged to be NC among those in a random sample of size n. For convenience, introduce $\mathbf{Y} \equiv (Y_1, \ldots, Y_k)$ and $\mathbf{N} \equiv (N_1, \ldots, N_k)$.

The distribution of $\mathbf{Y}$ is multi-hypergeometric, or symbolically

$$\mathbf{Y} \sim \text{Mult Hypg}(n; \mathbf{N}; N). \tag{2.19a}$$

Noting that

$$\binom{N-\sum N_i}{n-\sum y_i}=\binom{0}{0}=1,$$

the distribution of **Y** can be written as

$$\Pr[\mathbf{Y}=\mathbf{y}]=\left\{\prod_{i=1}^{k}\binom{N_i}{y_i}\right\}\binom{N-\sum N_i}{n-\sum y_i}\bigg/\binom{N}{n}$$

$$=\left\{\prod_{i=1}^{k}\binom{N_i}{y_i}\right\}\bigg/\binom{N}{n}, \tag{2.19b}$$

where $0 \leqslant y_i \leqslant N_i$. Given **Y**, the corresponding numbers of items in the k strata judged to be NC will be random variables which conditional distributions

$$\text{Bin}(Y_1, p_1), \ldots, \text{Bin}(Y_k, p_k),$$

respectively. Symbolically, using multifold convolution, we have the conditional distribution of Z, given **Y**:

$$Z|\mathbf{Y} \sim * \cdots *_{i=1}^{k} \text{Bin}(Y_i, p_i) \tag{2.20}$$

with, as above, $*$ denoting convolution and $* \cdots *^k$ a k-fold convolution. Compounding with respect to **Y**, we obtain the overall distribution of Z,

$$Z \sim [* \cdots *_{i=1}^{k} \text{Bin}(Y_i, p_i)] \underset{Y}{\wedge} \text{Mult Hypg}(n; \mathbf{N}; N). \tag{2.21}$$

Moments

Recall that the sth descending factorial moment of a $\text{Bin}(n, p)$ variable, X, say, is

$$\mu_{(s)} = E[X^{(s)}] = n^{(s)} p^s. \qquad [\text{cf. (2.4a)}]$$

Conditioned on **Y**, the sth factorial moment of Z, a sum of k independent binomial variables, is

$$\mu_{(s)}(Z|\mathbf{Y}) = \sum_{j_1}\cdots\sum_{j_k}\binom{s}{j_1, \ldots, j_k}\prod_{i=1}^{k}(Y_i^{(j_i)} p_i^{j_i}) \qquad \left(j_i \geqslant 0;\ \sum_{i=1}^{k} j_i = s\right), \tag{2.22}$$

where $\binom{s}{j_1, \ldots, j_k} = s!/(\prod_{i=1}^{k} j_i!)$ is a multinomial coefficient.

Taking expectation with respect to **Y**, we have

$$\mu_{(s)}(Z) = \frac{n^{(s)}}{N^{(s)}} \sum_{j_1} \cdots \sum_{j_k} \binom{s}{j_1, \ldots, j_k} \prod_{i=1}^{k} (N_i^{(j_i)} p_i^{j_i}). \tag{2.23}$$

Note that the sth factorial moment for **Y** with distribution given by (2.19a) is:

$$\mu_{(\mathbf{s})}(\mathbf{Y}) = E[\mathbf{Y}^{(\mathbf{s})}] = n^{(s)} \left\{ \prod_{i=1}^{k} N_i^{(s_i)} \right\} \Big/ N^{(s)},$$

where $s = \sum_{i=1}^{k} s_i$. As a special case of (2.23) with $s = 1$,

$$E[Z] = nN^{-1} \sum_{i=1}^{k} N_i p_i = n\bar{p}, \tag{2.24}$$

where $\bar{p} = N^{-1} \sum_{i=1}^{k} N_i p_i$ is the probability of classifying a *randomly chosen* item as NC. Similarly, from (2.23), we also have

$$\text{var}(Z) = n\left(\frac{N-n}{N-1}\right)\bar{p}(1-\bar{p}) + \frac{n(n-1)}{N-1} \sum_{i=1}^{k} \frac{N_i}{N} p_i(1-p_i),$$

$$= n\bar{p}(1-\bar{p}) - \frac{n(n-1)}{N(N-1)} \sum_{i=1}^{N} N_i(p_i - \bar{p})^2. \tag{2.25}$$

Note that var(Z) is usually less than the 'binomial' value $n\bar{p}(1-\bar{p})$, but is greater than the hypergeometric value $n((N-n)/(N-1))\bar{p}(1-p)$.

Taking the special case $k = 2$, with $N_1 = D$ and $N_2 = N - D$, we obtain, from (2.24)

$$E[Z] = n\bar{p},$$

and from (2.25)

$$\text{var}(Z) = n\left(\frac{N-n}{N-1}\right)\bar{p}(1-\bar{p})$$

$$+ \frac{n(n-1)}{N}\left[\frac{D}{N} p_1(1-p_1) + \left(1 - \frac{D}{N}\right) p_2(1-p_2)\right]$$

$$= n\bar{p}(1-\bar{p}) - \frac{n(n-1)}{N-1} \frac{D}{N}\left(1 - \frac{D}{N}\right)(p_1 - p_2)^2,$$

with $\bar{p} = (D/N)p_1 + [1 - (D/N)]p_2$, which coincide with formulas (2.17) and (2.18), respectively. Further theoretical details are given in Kotz and Johnson (1983) and Johnson and Kotz (1983).

Waiting time

Formula (2.9) also applies to the situations described in this section and in 2.3, using the appropriate distributions (2.3a) or (2.20) of Z, with n replaced by g. Table 2.2 gives some illustrative numerical values of

$$\Pr[G=g] \qquad \text{and} \qquad \Pr[G=g]/\Pr[G \leqslant n].$$

for lot size 100 and sample size $n=10$, for $k=2$, as in section 2.3.

The second quantity ('Norm') is the normalized value for the conditional distribution of G, given that at least m NC decisions would be obtained from inspection of the random sample of n items. From the table one can assess the effects of changes in D, m, p and p'. As is to be expected, the required number of inspections (G) increases with m and decreases with D, p and p'.

In order to obtain expressions for the moments of G, we consider the ordered results of testing every one of the N items in the lot. Given Z, the total number of NC decisions, there are $\binom{N}{Z}$ possible orderings of the results. The probability that the mth NC decision is the gth decision in order is

$$\Pr[G=g]=\binom{N}{Z}^{-1}\binom{g-1}{m-1}\binom{N-g}{Z-m}$$

(a *negative* hypergeometric distribution with parameters m and N).

Given $Z(\geqslant m)$, the expected value of

$$G^{[r]}=G(G+1)\cdots(G+r-1)$$

(the rth **ascending factorial moment**) is

$$E[G^{[r]}|Z]=m^{[r]}(N+1)^{[r]}/(Z+1)^{[r]}. \tag{2.27}$$

To obtain the overall rth ascending moment of G, (2.27) has to be averaged over the distribution (2.20) of Z, truncated by $Z \geqslant m$. (If $Z<m$ there will not be m NC decisions, even though all N items in the lot are inspected.)

If the possibility that $Z<m$ is neglected, the following approximate values are obtained after approximating the expected values of $1/(Z+1)$ and $1/(Z+1)^{[2]}$:

$$E[G] \doteq m\bar{p}^{-1}\left[1+\frac{1}{N}\left\{1+\frac{1}{\bar{p}}+\frac{1}{\bar{p}^2 N}\sum_{j=1}^{k} N_j p_j(1-p_j)\right\}\right]; \tag{2.28a}$$

Table 2.2 *Distributions of waiting time until m items declared* NC ($k = 2$); *Lot size* (N) = 100, *sample size* (n) = 10

	$D = 5$				$D = 10$			
	$m = 1$		$m = 2$		$m = 1$		$m = 2$	
$p = 0{\cdot}75$; $p' = 0{\cdot}05$								
g	Pr [$G = g$]	Norm	Pr [$G = g$]	Norm	Pr [$G = g$]	Norm	Pr [$G = g$]	Norm
1	0·0850	0·1431	–	–	0·1200	0·1646	–	–
2	0·0780	0·1313	0·0070	0·0341	0·1060	0·1455	0·0140	0·0408
3	0·0716	0·1205	0·0129	0·0628	0·0937	0·1285	0·0248	0·0724
4	0·0657	0·1105	0·0178	0·0866	0·0827	0·1134	0·0329	0·0963
5	0·0602	0·1014	0·0218	0·1063	0·0729	0·1001	0·0389	0·1138
6	0·0552	0·0929	0·0250	0·1222	0·0643	0·0882	0·0431	0·1260
7	0·0506	0·0852	0·0276	0·1348	0·0567	0·0778	0·0458	0·1339
8	0·0464	0·0781	0.0296	0·1445	0·0499	0·0685	0·0473	0·1382
9	0·0425	0·0715	0·0311	0·1518	0·0440	0·0603	0·0478	0·1397
10	0·0389	0·0655	0·0322	0·1569	0·0387	0·0530	0·0476	0·1390
$p = 0{\cdot}95$; $p' = 0{\cdot}05$								
g								
1	0·0950	0·1485	–	–	0·1400	0·1775	–	–
2	0·0865	0·1350	0·0086	0·0355	0·1211	0·1535	0·0189	0·0448
3	0·0785	0·1227	0·0158	0·0649	0·1047	0·1327	0·0328	0·0779
4	0·0713	0·1114	0·0216	0·0888	0·0904	0·1146	0·0428	0·1017
5	0·0647	0·1012	0·0262	0·1080	0·0780	0·0989	0·0496	0·1178
6	0·0587	0·0918	0·0299	0·1235	0·0673	0·0853	0·0538	0·1278
7	0·0533	0·0833	0·0327	0·1347	0·0597	0·0734	0·0560	0·1330
8	0·0483	0·0755	0·0348	0·1432	0·0498	0·0632	0·0567	0·1345
9	0·0438	0·0685	0·0362	0·1490	0·0428	0·0543	0·0561	0·1331
10	0·0397	0·0620	0·0371	0·1526	0·0368	0·0466	0·0546	0·1295

	$m = 3$		$m = 4$		$m = 3$		$m = 4$	
$p = 0{\cdot}75;\ p' = 0{\cdot}05$								
g								
1	0·0000	0·0000	0·0000	0·0000	0·0000	0·0000	0·0000	0·0000
2	0·0000	0·0000	0·0000	0·0000	0·0000	0·0000	0·0000	0·0000
3	0·0006	0·0126	0·0000	0·0000	0·0016	0·0151	0·0000	0·0000
4	0·0015	0·0349	0·0000	0·0069	0·0042	0·0403	0·0002	0·0081
5	0·0028	0·0645	0·0002	0·0256	0·0075	0·0717	0·0006	0·0289
6	0·0044	0·0991	0·0004	0·0591	0·0111	0·1063	0·0014	0·0646
7	0·0060	0·1371	0·0007	0·1092	0·0148	0·1419	0·0025	0·1155
8	0·0078	0·1769	0·0011	0·1766	0·0184	0·1766	0·0038	0·1804
9	0·0096	0·2174	0·0016	0·2610	0·0218	0·2092	0·0055	0·2577
10	0·0114	0·2575	0·0023	0·3617	0·0249	0·2390	0·0073	0·3488
$p = 0{\cdot}95;\ p' = 0{\cdot}05$								
g								
1	0·0000	0·0000	0·0000	0·0000	0·0000	0·0000	0·0000	0·0000
2	0·0000	0·0000	0·0000	0·0000	0·0000	0·0000	0·0000	0·0000
3	0·0008	0·0131	0·0000	0·0000	0·0024	0·0165	0·0000	0·0000
4	0·0021	0·0360	0·0001	0·0071	0·0064	0·0432	0·0003	0·0087
5	0·0038	0·0659	0·0002	0·0261	0·0112	0·0756	0·0011	0·0307
6	0·0058	0·1006	0·0005	0·0600	0·0164	0·1101	0·0024	0·0675
7	0·0079	0·1381	0·0010	0·1103	0·0214	0·1442	0·0041	0·1186
8	0·0102	0·1769	0·0016	0·1773	0·0262	0·1762	0·0064	0·1822
9	0·0124	0·2158	0·0023	0·2605	0·0304	0·2048	0·0089	0·2558
10	0·0146	0·2537	0·0032	0·3586	0·0341	0·2294	0·0118	0·3365

$$\operatorname{var}(G) \doteq \frac{m(1-\bar{p})}{\bar{p}^2}\left[1+\frac{1}{N}-\frac{3+m}{N\bar{p}}\right.$$
$$\left.+\frac{3+m-(m+1)\bar{p}}{N\bar{p}^2(1-\bar{p})}\sum_{j=1}^{k} N_j p_j(1-p_j)\right]. \qquad (2.28b)$$

Details of the derivation are given in Johnson and Kotz (1983).

It can be shown that this (approximate) formula gives values of var(G) between

$$\frac{m(1-\bar{p})}{\bar{p}^2}\left[1+\frac{1}{N}-\frac{3+m}{N\bar{p}}\right] \quad \text{and} \quad \frac{m(1-\bar{p})}{\bar{p}^2}\left(1+\frac{1}{N}\right)$$

(Johnson and Kotz (1983)).

CHAPTER 3

Basic distributions: multiple sampling

3.1 Introduction

As in Chapter 2, we will consider a population (lot) of size N, containing D NC items. However, instead of taking a single sample, we will take k, say, random samples of sizes $\mathbf{n} \equiv (n_1, \ldots, n_k)$, respectively, without replacement. Let $\mathbf{Y} \equiv (Y_1, \ldots, Y_k)$ be the numbers of truly NC items in the corresponding samples, with $\mathbf{y} \equiv (y_1, \ldots, y_k)$ as a realized value of Y.

For the jth sample, we assume the probability of correctly (incorrectly) declaring an NC (C) item as NC on inspection is p_j (p'_j) for $j = 1, \ldots, k$.

We will study the distribution of the numbers $\mathbf{Z} \equiv (Z_1, \ldots, Z_k)$ of items declared NC after inspection. As before, we will also obtain expressions for relevant moments.

In section 3.2, the special case of equal error probabilities for all samples is considered while more general cases with unequal error probabilities will be discussed in section 3.3.

3.2 The case of equal error probabilities

Here we assume the error probabilities are the same for all samples, so that $p = p_1 \cdots = p_k$ and $p' = p'_1 \cdots = p'_k$.

Distributions

Let $\mathbf{Y} \equiv (Y_1, \ldots, Y_k)$, where Y_j is the (unknown) number of truly NC items in the jth sample, and $\mathbf{y} \equiv (y_1, \ldots, y_k)$, a realized value of $\mathbf{Y}$. We use the symbol

$$\binom{a}{\mathbf{b}} \equiv \binom{a}{b_1, \ldots, b_k} = \frac{a!}{(a - \sum_{j=1}^{k} b_j)! \prod_{j=1}^{k} b_j!}$$

to denote a multinomial coefficient.

The joint distribution of Y is multivariate hypergeometric with parameters $(\mathbf{n}; D, N)$:

$$\Pr[\mathbf{Y}=\mathbf{y}] = \binom{D}{\mathbf{y}}\binom{N-D}{\mathbf{n}-\mathbf{y}}\bigg/\binom{\mathbf{N}}{\mathbf{n}}$$

$$= \binom{N}{D}^{-1}\left\{\prod_{i=1}^{k}\binom{n_i}{y_i}\right\}\binom{n-\sum n_i}{D-\sum y_i}, \tag{3.1a}$$

with the following constraints on $\mathbf{y} = (y_1, y_2, \ldots, y_k)$:

$$0 \leqslant y_i \leqslant n_i\ (i=1,\ldots,k) \qquad \text{and} \qquad D - N + \sum_{i=1}^{k} n_i \leqslant \sum_{i=1}^{k} y_i.$$

Symbolically,

$$\mathbf{Y} \sim \text{Mult Hypg}(\mathbf{n}; D, N). \tag{3.1b}$$

Let $\mathbf{Z} \equiv (Z_1, \ldots, Z_k)$, where Z_j is the number of items in the jth sample classified as NC. Conditional on $\mathbf{Y}$, the Z's are mutually independent and, as in the case of single sampling, each has a distribution of form

$$Z_i | Y_i \sim \text{Bin}(Y_i, p) * \text{Bin}(n_i - Y_i, p'). \tag{3.2}$$

The joint conditional distribution of $\mathbf{Z}$, given $\mathbf{Y}$, can be written compactly in vector form

$$\mathbf{Z}|\mathbf{Y} \sim \text{Bin}(\mathbf{Y}, p) * \text{Bin}(\mathbf{n} - \mathbf{Y}, p'). \tag{3.3}$$

where

$$\text{Bin}(\mathbf{Y}, p) * \text{Bin}(\mathbf{n} - \mathbf{Y}, p') \equiv \begin{pmatrix} \text{Bin}(Y_1, p) * \text{Bin}(n_1 - Y_1, p') \\ \vdots \quad \vdots \quad \vdots \\ \text{Bin}(Y_k, p) * \text{Bin}(n_k - Y_k, p') \end{pmatrix}.$$

The overall distribution of $\mathbf{Z}$ is then

$$\mathbf{Z} \sim \{\text{Bin}(\mathbf{Y}, p) * \text{Bin}(\mathbf{n} - \mathbf{Y}, p')\} \bigwedge_{\mathbf{Y}} \text{Mult Hypg}(\mathbf{n}, D, N). \tag{3.4a}$$

Explicitly,

$$\Pr[\mathbf{Z}=\mathbf{z}|\mathbf{n}; D, N; p, p'] = \binom{N}{\mathbf{n}}^{-1} \sum_{\mathbf{y}} \binom{D}{\mathbf{y}}\binom{N-D}{\mathbf{n}-\mathbf{y}}$$

$$\times \left[\prod_{j=1}^{k}\left\{\sum_{w_j}\binom{y_j}{w_j} p^{w_j}(1-p)^{y_j-w_j}\right.\right.$$

$$\left.\left.\times \binom{n_j - y_j}{z_j - w_j} p'^{z_j - w_j}(1-p')^{n_j - y_j - z_j + w_j}\right\}\right]. \tag{3.4b}$$

Moments

The $(r_1,\ldots,r_k)$ mixed factorial moment of $\mathbf{Z}$, conditioned on $\mathbf{Y}$, using (2.18) of Chapter 2, is

$$E\left[\prod_{j=1}^{k} Z_j^{(r_j)} \mid \mathbf{Y}\right] = \prod_{j=1}^{k} E[Z_j^{(r_j)} \mid Y_j]$$

$$= \prod_{j=1}^{k} \left\{ \sum_{w_j=0}^{r_j} \binom{r_j}{w_j} p^{w_j} p'^{r_j - w_j} Y_j^{(w_j)} (n_j - Y_j)^{(r_j - w_j)} \right\}. \tag{3.5}$$

From (2.4c), we obtain

$$E_{\mathbf{Y}}\left[\prod_{j=1}^{k} \left\{ Y_j^{(\alpha_j)} (n_j - Y_j)^{(\beta_j)} \right\}\right] = \left\{ \prod_{j=1}^{k} n_j^{(\alpha_j + \beta_j)} \right\} D^{(\alpha)} (N - D)^{(\beta)} / N^{(\alpha + \beta)}. \tag{3.6}$$

where $\alpha = \sum_{j=1}^{k} \alpha_j$ and $\beta = \sum_{j=1}^{k} \beta_j$. The unconditional $\mathbf{r}$th factorial moment of $\mathbf{Z}$ is therefore

$$\mu_{(\mathbf{r})}(\mathbf{Z}) \equiv E\left[\prod_{j=1}^{k} Z_j^{(r_j)}\right]$$

$$= \left\{ \prod_{j=1}^{k} n_j^{(r_j)} \right\} \sum_{w=0}^{r} \binom{r}{w} p^w p'^{r-w} D^{(w)} (N - D)^{(w)} / N^{(r)}, \tag{3.7}$$

where $\mathbf{r} \equiv (r_1,\ldots,r_k)$ and $r = \sum_{j=1}^{k} r_j$.

As a special case of (3.7), we obtain the covariance

$$\operatorname{cov}(Z_j, Z_{j'}) = E[Z_j Z_{j'}] - E[Z_j]E[Z_{j'}]$$

$$= \frac{n_j n_{j'}}{N(N-1)} (N^2 \bar{p}^2 - N \overline{p^2}) - n_j n_{j'} \bar{p}^2$$

$$= -n_j n_{j'} (N-1)^{-1} (\overline{p^2} - \bar{p}^2), \tag{3.8}$$

where (as in Chapter 2)

$$\bar{p} = \frac{Dp + (N - D)p'}{N}, \qquad \text{and} \qquad \overline{p^2} = \frac{Dp^2 + (N - D)p'^2}{N}.$$

From (3.8) we see that $\operatorname{cov}(Z_j, Z_{j'}) \leqslant 0$, because $\overline{p^2} \geqslant \bar{p}^2$. This is understandable: since the total number of NC items is fixed, if one sample has a greater number of NC items, then others must have fewer, and this is reflected by the observed values of the Z's, provided the inspection is reasonable.

Remarks

It can be seen that the multi-hypergeometric distribution obtained in this section for $\mathbf{Y}$ is the same as the one obtained in section 2.4 in another context. In section 2.4, instead of taking k random samples, we were dealing with a lot of N items, divided into k strata, but selecting a single sample, leading to the same distribution. This demonstrates that the same mathematical structure can arise from a number of different statistical models, with different explanations.

The techniques used in this section and the following section, are in some respects analogous to those used in Chapter 2.

3.3 The case of unequal error probabilities

Distributions

Here we consider the same setting as in section 3.2 except that the error probabilities may vary from sample to sample. Of course, $\mathbf{Y}$ has the same distribution as in section 3.2 (see (3.1a) and (3.1b). Given Y_j, Z_j is the sum of two mutually independent binomially distributed variables (with parameters (Y_j, p_j) and $(n_j - Y_j, p'_j)$, respectively), which are

W_j: the number of NC items in the jth sample classified as NC;
W'_j: the number of C items in the jth sample classified as NC, i.e. $Z_j = W_j + W'_j$ for $j = 1,\ldots,k$. Conditional on $\mathbf{Y}$, the Z's are mutually independent.

We therefore have

$$Z_i | Y_i \sim \text{Bin}(Y_i, p_i) * \text{Bin}(n_i - Y_i, p'_i). \tag{3.9}$$

The joint conditional distribution of $\mathbf{Z}$ given $\mathbf{Y}$ can be written in vector form

$$\mathbf{Z}|\mathbf{Y} \sim \text{Bin}(\mathbf{Y}, \mathbf{p}) * \text{Bin}(\mathbf{n} - \mathbf{Y}, \mathbf{p}'), \tag{3.10a}$$

where, with $\mathbf{p} = (p_1,\ldots,p_k)$ and $\mathbf{p}' = (p'_1,\ldots,p'_k)$,

$$\text{Bin}(\mathbf{Y}, \mathbf{y}) * \text{Bin}(\mathbf{n} - \mathbf{Y}, \mathbf{p}') \equiv \begin{pmatrix} \text{Bin}(Y_1, p_1) * \text{Bin}(n_1 - Y_1, p'_1) \\ \vdots \quad\quad \vdots \quad\quad \vdots \\ \text{Bin}(Y_k, p_k) * \text{Bin}(n_k - Y_k, p'_k) \end{pmatrix}.$$

Explicitly,

$$\begin{aligned}\Pr[\mathbf{Z}=\mathbf{z}\mid\mathbf{Y}=\mathbf{y}]\\ &=\prod_{i=1}^{k}\left[\prod_{h=0}^{z_i}\binom{y_i}{h}\binom{n_i-y_i}{z_i-h}\right.\\ &\qquad\left.\times\, p_i^h p_i'^{z_i-h}(1-p_i)^{y_i-h}(1-p_i')^{n_i-y_i-z_i+h}\right]\\ &=\prod_{i=1}^{k} b(z_i;\, y_i, p_i;\, n_i-y_i, p_i'),\end{aligned} \tag{3.10b}$$

where $b(z_i; y_i, p_i; n_i - y_i, p')$, defined by (3.10b), is the probability function of the convolution (sum) of the two independent binomial variables, with parameters (y_i, p_i) and $(n_i - y_i, p_i')$, respectively.

Compounding the distribution (3.10a,b) by the distribution (3.1a,b), the overall distribution of $\mathbf{Z}$ is, symbolically,

$$\mathbf{Z}\sim\begin{pmatrix}\mathrm{Bin}(Y_1,p_1)*\mathrm{Bin}(n_1-Y_1,p_1')\\ \vdots \qquad\qquad \vdots\\ \mathrm{Bin}(Y_k,p_k)*\mathrm{Bin}(n_k-Y_k,p_k')\end{pmatrix}\underset{\mathbf{Y}}{\wedge}\text{Mult Hypg}_k(\mathbf{n};D,N), \tag{3.11a}$$

a **multivariate hypergeometric–convoluted binomial(s) distribution**, or explicitly,

$$\begin{aligned}\Pr[\mathbf{Z}=\mathbf{z}]=\binom{N}{n}^{-1}\sum_{y_1}\cdots\sum_{y_k}\left\{\binom{N-\sum N_i}{D-\sum y_i}\prod_{i=1}^{k}\binom{n_i}{y_i}\right\}\\ \times\prod_{i=1}^{k} b(z_i;\, y_i, p_i;\, n_i-y_i, p_i').\end{aligned} \tag{3.11b}$$

Here the limits for $\mathbf{y}$ are given by

$$0\leqslant y_i\leqslant n_i (i=1,\ldots,k) \quad\text{and}\quad D-N+\textstyle\sum_{i=1}^{k} n_i\leqslant\sum_{i=1}^{k} y_i.$$

Moments

From (2.15) of Chapter 2, the r_ith factorial descending moment of Z_i, conditioned on $\mathbf{Y}$, is

$$E[Z_i^{(r_i)}\mid\mathbf{Y}]=\sum_{h=0}^{r_i}\binom{r_i}{h}p_i^h p_i'^{r_i-h}Y_i^{(h)}(n_i-Y_i)^{(r_i-h)}. \tag{3.12}$$

Taking expectation with respect to $\mathbf{Y}$ and applying (2.4c), the overall

$\mathbf{r} = (r_1, \ldots, r_k)$th factorial moment of $\mathbf{Z}$ is

$$\mu_{(\mathbf{r})}(\mathbf{Z}) = E\left[\prod_{i=1}^{k} Z_i^{(r_i)}\right] = E_{\mathbf{Y}}E\left[\prod_{i=1}^{k} Z_i^{(r_i)}|\mathbf{Y}\right] = E_{\mathbf{Y}}\left\{\prod_{i=1}^{k} E[Z_i^{(r_j)}|\mathbf{Y}]\right\}$$

$$= E\left[\prod_{i=1}^{k}\left\{\sum_{h=0}^{r_i}\binom{r_i}{h}p_i^h p_i'^{r_i-h} Y_i^{(h)}(n_i - Y_i)^{(r_i-h)}\right\}\right]$$

$$= \sum_{l_1=0}^{r_1}\cdots\sum_{l_k=0}^{r_k}\left\{\prod_{i=1}^{k}\binom{r_i}{l_i}p_i^{l_i}p_i'^{r_j-l_i}E[Y_i^{(l_i)}(n_i - Y_i)^{(r_i-h)}]\right\}$$

$$= \frac{\prod_{i=1}^{k} n_i^{(r_i)}}{N^{(\Sigma r_i)}}\sum_{l_1=0}^{r_1}\cdots\sum_{l_k=0}^{r_k}\left\{\prod_{i=1}^{k}\binom{r_i}{l_j}p_i^{l_i}p_i'^{r_i-l_i}\right\}D^{(\Sigma l_i)}(N-D)^{(\Sigma(r_i-l_i))}$$

$$= \frac{\prod_{i=1}^{k} n^{(r_i)}}{N^{(\Sigma r_i)}}\sum_{l=0}^{\Sigma r_i} D^{(l)}(N-D)^{(r-l)}g_l, \tag{3.13}$$

where g_l is the coefficient of x^l in $\prod_{i=1}^{k}(p_i' + p_i x)^{r_i}$.

In spite of its rather formidable appearance formula (3.13) is a straightforward generalization of the corresponding formulas for the factorial moment of Z for the case of single sampling derived in Chapter 2.

Remarks

Expression (3.13) can be specialized in several ways. We note the following.

1. For sampling with replacement, the corresponding factorial moment

$$\mu_{(\mathbf{r})}(\mathbf{Z}) = \left\{\prod_{i=1}^{k} n_i^{(r_i)}\right\}\prod_{i=1}^{k}\{Dp_i + (N-D)p_i'\}^{r_i}\Bigg]\Bigg/ N^{\Sigma r_i}. \qquad \text{(cf. 3.7))}$$

2. For the equal error probability case, i.e. $p_i = p$ and $p_i' = p'$ for all i,

$$g_l = \binom{\Sigma r_i}{l}p^l p'^{r-l}.$$

3. With $r = 1$, as in (2.24), we have the mean value

$$E[Z_i] = n_i[DN^{-1}p_i + (1 - DN^{-1})P_i'] = n_i\bar{p}_i \tag{3.14}$$

where $\bar{p}_i = DN^{-1}p_i + (1 - DN^{-1})p_i'$, the probability that an item in the ith sample will be judged as NC, either correctly or incorrectly. We also obtain, similar to (2.25),

$$\mathrm{var}(Z_i) = n_i\bar{p}_i(1-\bar{p}_i) - \frac{n_i(n_i-1)}{(N-1)}\frac{D}{N}\left(1-\frac{D}{N}\right)(p_i - p_i')^2 \qquad (3.15)$$

and also

$$\mathrm{cov}(Z_i, Z_j) = -n_i n_j (N-1)^{-1} D N^{-1}(1 - DN^{-1})(p_i - p_i')(p_j - p_j'). \qquad (3.16)$$

It is clear that

$$\mathrm{cov}(Z_i, Z_j)\begin{cases} <0 & \text{if } p_i > p_i' \text{ and } p_j > p_j'; \\ =0 & \text{if } p_i = p_i' \text{ or } p_j = p_j'. \end{cases}$$

In practice, usually $p_i > p_i'$ and $p_j > p_j'$, otherwise the inspection would be useless. Therefore, the above covariance is negative. An intuitive explanation is that, since the covariance of Y_i and Y_j is negative, (because the total number of NC items is fixed), so is that of Z_i and Z_j, which are 'observations' of Y_i and Y_j, provided the inspection is reasonable. The condition $p_i = p_i'$ means that any item, NC or not, from the ith sample, whether NC or C, has the same chance to be declared NC, and therefore the result of inspection (as exposed in Z_i) is independent of Y_i or the composition of the sample. Hence the covariance should be zero, reflecting the independence between the corresponding Z (or set of Z's) and all other Z's.

Regression function

Since the conditional distribution of Z_i, given Z_j, can be obtained straightforwardly, but has a cumbersome complex form, we prefer to derive directly the following useful regression functions.

We have, from (3.12),

$$E[Z_i \mid Y_i] = Y_i p_i + (n_i - Y_i)p_i',$$

so, by taking expectation with respect to Y_i conditioned on Z_j, $i \neq j$,

$$E[Z_i|Z_j] = p_i E[Y_i|Z_j] + p_i'(n_i - E[Y_i|Z_j]). \qquad (3.17)$$

Now, noting the simple fact that (recall the distribution of $\mathbf{Y}$, given by (3.1a)):

$$E[Y_i|Y_j]=\frac{n_i(D-Y_j)}{N-n_j},$$

and taking expectation with respect to Y_j conditioned on Z_j, we arrive at

$$E[Y_i|Z_j]=\frac{n_i(D-E[Y_j|Z_j])}{N-n_j},\qquad i\neq j. \tag{3.18}$$

From (3.17) in order to obtain $E[Z_i|Z_j]$, we only need to evaluate $E[Y_j|Z_j]$. For the jth sample,

$$\Pr[\text{item is NC}|\text{ item classified as NC}]=\frac{D}{N}\frac{p_j}{\bar{p}_j}$$

where $\bar{p}_j=DN^{-1}p_j+(1-DN^{-1})p_j'$, as above. Also

$$\Pr[\text{item is NC}|\text{ item classified as C}]=\frac{D}{N}\frac{(1-p_j)}{(1-\bar{p}_j)}.$$

Hence

$$E[Y_j|Z_j]=\frac{D}{N}\left[Z_j\frac{p_j}{\bar{p}_j}+(n_j-Z_j)\frac{1-p_j}{1-\bar{p}_j}\right]. \tag{3.19}$$

From formulas (3.17) and (3.18), we have

$$\begin{aligned}E[Z_i|Z_j]&=n_ip_i'+(p_i-p_i')E[Y_i|Z_j]\\&=n_ip_i'+n_i(p_i-p_i')\frac{D-E[Y_j|Z_j]}{N-n_j}\\&=n_ip_i'+n_i(p_i-p_i')\frac{D}{N}\left\{1-\frac{\left(1-\dfrac{D}{N}\right)(p_j-p_j')}{(N-n_j)\bar{p}_j(1-\bar{p}_j)}(Z_j-n_j\bar{p}_j)\right\}\end{aligned}$$

$$i\neq j. \tag{3.20}$$

The regression is linear, and the sign of the regression coefficient is the sign of $(p_i-p_i')(p_j'-p_j)$, as is that of $\operatorname{cov}(Z_i,Z_j)$ in (3.16).

A special case

If each sample has only one item, i.e. $n_i=1$ for $i=1,\ldots,k$, each one of the Y's and Z's is either 0 or 1. The distribution (3.11b) then becomes

$$\Pr[\mathbf{Z}=\mathbf{z}] = \binom{N}{D}^{-1} \sum_{y_1} \cdots \sum_{y_k} \binom{N-k}{D-\sum_{i=1}^{k} y_i} \times \prod_{i=1}^{k} \{y_i p_i^{z_i}(1-p_i)^{1-z_i} + (1-y_i)p_i'^{z_i}(1-p_i')^{1-z_i}\}$$

$$(z_i = 0, 1;\, y_i = 0, 1). \quad (3.21)$$

The total number of NC items in the k samples is $\sum y_i = y$. Collecting the terms with the same value of y together in (3.21) yields an alternative expression. Let

$$(\boldsymbol{\alpha}, \boldsymbol{\beta}) \equiv \{(\alpha_i, \alpha_i', \beta_i, \beta_i')\}.$$

We have

$$\Pr(\mathbf{Z}=\mathbf{z}] = \binom{N}{D}^{-1} \sum_{y} \binom{N-k}{D-y} \sum_{(\boldsymbol{\alpha},\boldsymbol{\beta})} \prod_{i=1}^{k} p_i^{\alpha_i} p_i'^{\alpha_i'} (1-p_i)^{\beta_i} (1-p_i')^{\beta_i'}, \quad (3.21a)$$

where the summation over $(\boldsymbol{\alpha}, \boldsymbol{\beta})$ is constrained by

$$\sum_{i=1}^{k} (\alpha_i + \beta_i) = y \quad \text{and} \quad \sum_{i=1}^{k} (\alpha_i' + \beta_i') = \sum_{i=1}^{k} z_i,$$

and, for each i, one of $(\alpha_i, \alpha_i', \beta_i, \beta_i')$ is 1, the other three are all zero.

Yet a third form of (3.21) is

$$\Pr(\mathbf{Z}=\mathbf{z}] = \binom{N}{D}^{-1} \sum_{y} \binom{N-k}{D-y} \left[\text{coefficient of } x^z u^y \text{ in} \prod_{j=1}^{k} \{p_j xu + p_j' x + (1-p_j)u + 1 - p_j'\} \right]. \quad (3.21b)$$

Remark

Expression (3.21b) shows that $\sum_{i=1}^{k} Z_i$ is a sufficient statistic for estimating D, when N, p's are known. This is also evident from intuitive considerations, in view of the special (0, 1) support of Z_i in this case.

More details of theoretical derivations are given in Johnson and Kotz (1983, 1985b, c), and some notes on tables are contained in section 4.2.

CHAPTER 4

Double and link sampling for acceptance

4.1 Introduction

In this chapter, we are going to discuss procedures of acceptance based on double sampling and link sampling, in which the decision as to whether a lot is acceptable or not depends on results of inspection of one or more samples, and the number of samples taken from the lot is determined in part by the results of initial inspections.

In double sampling, the decision structure allows for taking up to two samples from the same lot before decision on acceptance is made. A link (or partial link) sampling procedure allows for using results of inspection of samples from different 'neighbouring' lots if no decision can be made after testing the first sample.

Technical details relating to the formulas used in this chapter can be found in Johnson, Kotz and Rodriguez (1984; 1986).

4.2 Double and multi-stage sampling for acceptance

Given a lot of size N, containing D NC items, we take a sample of size n and inspect each item in the sample with a view to deciding whether there is sufficient reason to reject or accept the lot. If the evidence for an immediate decision is insufficient, we will take a second sample. As in previous chapters, the probability of incorrect classification of an NC (or C) item will be denoted by $1 - p$ (or p'), respectively.

Our decision rule is defined by three constants a_1, a'_1 and a_2, where a_1 and $(a'_1 + 1)$ are the boundary values for acceptance and rejection, based on the results from inspection of the first sample while a_2 is the boundary value for acceptance after examining the second sample. Varying these three boundary values changes the

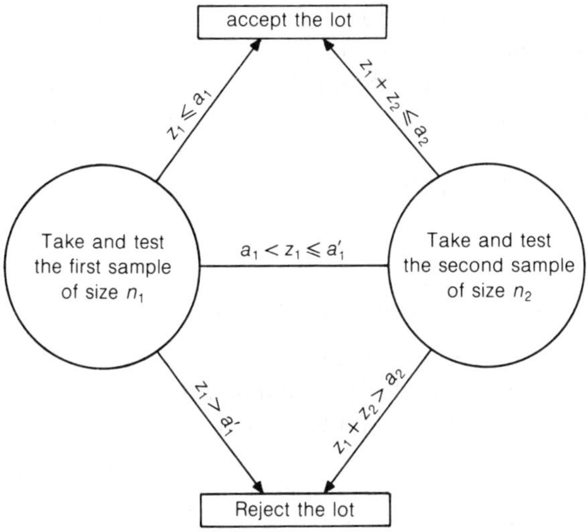

Figure 4.1 *Flow chart for double sampling*

probability of acceptance for a given proportion of NC items. Let n_1 and n_2 be the sample sizes in the first and second stages and let Z_1 and Z_2 be the numbers of items classified as NC in the first and second samples, respectively. All sampling is without replacement. Formally our procedure is as follows (and as illustrated in Fig. 4.1).

1. Take a random sample of size n_1, and obtain the number, Z_1, of items judged to be NC in the first sample.
2. If $Z_1 \leqslant a_1$, accept the lot. If $Z_1 > a'_1$, reject the lot.
3. If $a_1 < Z_1 \leqslant a'_1$, take a further sample of size n_2, and obtain the number, Z_2, of items judged NC in the second sample.
4. If $Z_1 + Z_2 \leqslant a_2$, accept the lot. If $Z_1 + Z_2 > a_2$, reject the lot.

The overall acceptance probability is

$$\Pr[Z_1 \leqslant a_1] + \Pr[(a_1 < Z_1 \leqslant a'_1) \cap (Z_1 + Z_2 \leqslant a_2)]. \quad (4.1)$$

Since our decision depends on the numbers Z_i ($i = 1, 2$) of declared NC items in the relevant sample(s), we need the distribution of Z_1 and the joint distribution of Z_1 and Z_2. The former one is given in (2.13b), and the latter is a special case of (3.4b) with $k = 2$.

The probabilities in (4.1) can be calculated from the expressions for the (joint) distribution of Z's for single and multiple sampling already derived in previous chapters. (We only need $k = 2$ here.) Specifically, the first probability in (4.1) can be obtained by summing (2.13b) over $Z_1 \leqslant a_1$; and the second is evaluated by summing (3.4b), taking $k = 2$, over the ranges $(a_1 < Z_1 \leqslant a'_1)$ and $(Z_1 + Z_2 \leqslant a_2)$. The calculations are somewhat tedious, although straightforward. Tables for the acceptance probabilities are therefore very convenient for practical use.

Tables and graphs of acceptance probabilities

In Table 3 and Figs. 3–6 of Johnson *et al.* (1986), acceptance probabilities are given for the double sampling procedure. These tables and figures correspond to the following sets of combinations

Table 4.1 *Acceptance probabilities: lot size* $N = 100$; AQL = 2·5; *sample sizes* $n_1 = 13$ *and* $n_2 = 13$; *acceptance numbers* $a_1 = 0$, $a'_1 = 1$, $a_2 = 1$

N (lot size) = 100 $D = 5$					
p	$p' = 0$	$p' = 0{\cdot}01$	$p' = 0{\cdot}02$	$p' = 0{\cdot}05$	$p' = 0{\cdot}10$
0·75	0·8023	0·7180	0·6368	0·4269	0·2015
0·90	0·7356	0·6514	0·5728	0·3771	0·1752
0·95	0·7127	0·6291	0·5518	0·3612	0·1669
0·98	0·6988	0·6158	0·5392	0·3518	0·1621
1·00	0·6895	0·6068	0·5309	0·3456	0·1590
N (lot size) = 100 $D = 10$					
p	$p' = 0$	$p' = 0{\cdot}01$	$p' = 0{\cdot}02$	$p' = 0{\cdot}05$	$p' = 0{\cdot}10$
0·75	0·4840	0·4233	0·3690	0·2402	0·1129
0·90	0·3765	0·3268	0·2829	0·1815	0·0841
0·95	0·3443	0·2983	0·2578	0·1647	0·0760
0·98	0·3260	0·2820	0·2435	0·1552	0·0715
1·00	0·3141	0·2716	0·2343	0·1492	0·0686
N (lot size) = 100 $D = 20$					
p	$p' = 0$	$p' = 0{\cdot}01$	$p' = 0{\cdot}02$	$p' = 0{\cdot}05$	$p' = 0{\cdot}10$
0·75	0·1355	0·1184	0·1033	0·0683	0·0337
0·90	0·0749	0·0652	0·0567	0·0372	0·0181
0·95	0·0610	0·0531	0·0461	0·0302	0·0146
0·98	0·0538	0·0468	0·0407	0·0266	0·0128
1·00	0·0495	0·0430	0·0374	0·0244	0·0117

of the parameters:

$$N = 100, 200;\ D/N = 0{\cdot}05.\ 0{\cdot}10,\ 0{\cdot}20;$$

$$p = 0{\cdot}75,\ 0{\cdot}90,\ 0{\cdot}95,\ 0{\cdot}98,\ 1{\cdot}00;\ p' = 0,\ 0{\cdot}01,\ 0{\cdot}02,\ 0{\cdot}05,\ 0{\cdot}10;$$

$$\text{AQL} = 1{\cdot}5\%,\ 2{\cdot}5\%\ 4\%,\ 6{\cdot}5\%,\ 10\%,$$

where AQL stands for acceptable quality level of the percentage of NC items. Some of these tables are reproduced here as Tables 4.1–4.4.

Using the tables and figures, one may compare the acceptance probabilities among different situations, ranging from perfect inspection with $p = 1$ and $p' = 0$ to cases involving large errors. For fixed p' the acceptance probability increases as p decreases, while for fixed p, the acceptance probability decreases as p' increases, since either large p or large p' tends to increase the number of items judged to

Table 4.2 *Acceptance probabilities: lot size* $N = 100$; AQL $= 4{\cdot}0$; *sample sizes* $n_1 = 13$ *and* $n_2 = 13$; *acceptance numbers* $a_1 = 0$, $a'_1 = 2$, $a_2 = 3$

N (lot size) $= 100$ $D = 5$

p	$p' = 0$	$p' = 0{\cdot}01$	$p' = 0{\cdot}02$	$p' = 0{\cdot}05$	$p' = 0{\cdot}10$
0·75	0·9900	0·9715	0·9440	0·8150	0·5278
0·90	0·9821	0·9567	0·9220	0·7756	0·4822
0·95	0·9787	0·9508	0·9137	0·7617	0·4671
0·98	0·9765	0·9470	0·9085	0·7532	0·4581
1·00	0·9749	0·9444	0·9049	0·7475	0·4521

N (lot size) $= 100$ $D = 10$

p	$p' = 0$	$p' = 0{\cdot}01$	$p' = 0{\cdot}02$	$p' = 0{\cdot}05$	$p' = 0{\cdot}10$
0·75	0·8777	0·8286	0·7753	0·6041	0·3444
0·90	0·8007	0·7429	0·6835	0·5080	0·2707
0·95	0·7712	0·7113	0·6508	0·4762	0·2481
0·98	0·7527	0·6918	0·6308	0·4573	0·2351
1·00	0·7400	0·6786	0·6174	0·4448	0·2266

N (lot size) $= 100$ $D = 20$

p	$p' = 0$	$p' = 0{\cdot}01$	$p' = 0{\cdot}02$	$p' = 0{\cdot}05$	$p' = 0{\cdot}10$
0·75	0·4130	0·3692	0·3285	0·2253	0·1109
0·90	0·2523	0·2207	0·1923	0·1244	0·0561
0·95	0·2084	0·1811	0·1568	0·0995	0·0437
0·98	0·1846	0·1598	0·1378	0·0865	0·0374
1·00	0·1698	0·1466	0·1261	0·0786	0·0336

Table 4.3 *Acceptance probabilities: lot size* $N = 100$; AQL = 6·5; *sample sizes* $n_1 = 13$ *and* $n_2 = 13$; *acceptance numbers* $a_1 = 1$, $a'_1 = 3$, $a_2 = 4$

N (lot size) = 100 *D* = 5

p	$p' = 0$	$p' = 0{\cdot}01$	$p' = 0{\cdot}02$	$p' = 0{\cdot}05$	$p' = 0{\cdot}10$
0·75	0·9996	0·9970	0·9911	0·9454	0·7666
0·90	0·9991	0·9948	0·9864	0·9289	0·7313
0·95	0·9989	0·9939	0·9845	0·9228	0·7192
0·98	0·9987	0·9933	0·9832	0·9189	0·7118
1·00	0·9986	0·9929	0·9824	0·9163	0·7069

N (lot size) = 100 *D* = 10

p	$p' = 0$	$p' = 0{\cdot}01$	$p' = 0{\cdot}02$	$p' = 0{\cdot}05$	$p' = 0{\cdot}10$
0·75	0·9750	0·9562	0·9319	0·8295	0·6005
0·90	0·9491	0·9209	0·8874	0·7622	0·5201
0·95	0·9375	0·9061	0·8696	0·7375	0·4934
0·98	0·9298	0·8964	0·8581	0·7222	0·4774
1·00	0·9243	0·8896	0·8502	0·7119	0·4668

N (lot size) = 100 *D* = 20

p	$p' = 0$	$p' = 0{\cdot}01$	$p' = 0{\cdot}02$	$p' = 0{\cdot}05$	$p' = 0{\cdot}10$
0·75	0·6840	0·6392	0·5945	0·4657	0·2858
0·90	0·5140	0·4706	0·4291	0·3177	0·1797
0·95	0·4576	0·4162	0·3771	0·2743	0·1512
0·98	0·4246	0·3847	0·3472	0·2500	0·1357
1·00	0·4030	0·3642	0·3279	0·2345	0·1261

be NC items (correctly or incorrectly), and, thus makes it easier to reject a lot.

Example 1

For analysing the sampling schemes in MIL-STD-105D, computed operating characteristic (OC) curves at the 5% and 10% points are needed, and the sensitivity of the OC curves can be assessed by using Table 3 and Figs. 3–6. As an example we take the procedure defined by the following parameter values:

$$n_1 = 13, n_2 = 13; \quad a_1 = 0, a'_1 = 2; \quad a_2 = 3; \quad \text{AQL} = 4\%,$$

(Inspection Level II). If $0{\cdot}95 \leqslant p \leqslant 1$ and $0 \leqslant p' \leqslant 0{\cdot}05$, limits for

Table 4.4 *Acceptance probabilities: lot size* $N = 100$; AQL = 10·0; *sample sizes* $n_1 = 13$ *and* $n_2 = 13$; *acceptance numbers* $a_1 = 2$, $a_1' = 4$, $a_2 = 6$

N (lot size) = 100 $D = 5$

p	$p' = 0$	$p' = 0{\cdot}01$	$p' = 0{\cdot}02$	$p' = 0{\cdot}05$	$p' = 0{\cdot}10$
0·75	1·0000	0·9999	0·9997	0·9954	0·9545
0·90	1·0000	0·9999	0·9995	0·9935	0·9436
0·95	1·0000	0·9998	0·9994	0·9928	0·9396
0·98	1·0000	0·9998	0·9993	0·9923	0·9371
1·00	1·0000	0·9998	0·9993	0·9920	0·9355

N (lot size) = 100 $D = 10$

p	$p' = 0$	$p' = 0{\cdot}01$	$p' = 0{\cdot}02$	$p' = 0{\cdot}05$	$p' = 0{\cdot}10$
0·75	0·9990	0·9974	0·9946	0·9751	0·8872
0·90	0·9973	0·9941	0·9891	0·9587	0·8446
0·95	0·9964	0·9925	0·9865	0·9517	0·8286
0·98	0·9958	0·9914	0·9847	0·9472	0·8185
1·00	0·9953	0·9906	0·9834	0·9440	0·8117

N (lot size) = 100 $D = 20$

p	$p' = 0$	$p' = 0{\cdot}01$	$p' = 0{\cdot}02$	$p' = 0{\cdot}05$	$p' = 0{\cdot}10$
0·75	0·9357	0·9163	0·8941	0·8123	0·6408
0·90	0·8589	0·8289	0·7967	0·6903	0·5016
0·95	0·8247	0·7915	0·7565	0·6443	0·4550
0·98	0·8021	0·7672	0·7307	0·6158	0·4275
1·00	0·7862	0·7502	0·7129	0·5965	0·4093

the variation in the probability of acceptance are shown below:

	(a)	(b)	(c)	(d)
$N = 100$	5%	0·7475	0·9749	0·9787
	10%	0·4448	0·7400	0·7712

and

	(a)	(b)	(c)	(d)
$N = 200$	5%	0·6816	0·9708	0·9761
	10%	0·3257	0·6639	0·7048,

where the columns headed (a), (b), (c) and (d) correspond to '$D/N \times 100\%$', 'Lower limit for acceptance probability ($p = 1{\cdot}00$ and $p' = 0{\cdot}05$)', 'Acceptance probability if error-free' and 'Upper limit for acceptance probability ($p = 0{\cdot}95$ and $p' = 0$)', respectively.

These tables show the acceptance probability is only slightly influenced by the lot size when D/N is small. Note that for small D (which corresponds to a small total number of NC items), p has little influence on the acceptance probability, since p can be effective only through D.

Example 2

For the same setting as Example 1, with $N = 100$ except that $p' \doteqdot 0{\cdot}02$ and p is unknown, how could we find the lower bound for p allowing at most a 0·04 increase in the probability of acceptance? We proceed as follows:

1. find the probability of acceptance for $p = 1{\cdot}0$ and $p' = 0{\cdot}02$;
2. add 0·04 to this value;
3. use linear interpolation to determine the value of p for $p' = 0{\cdot}02$ that corresponds to this probability of acceptance obtained from 1.

The results, as a function of D, are shown below:

D	Lower bound for p
5	0·75
10	0·94
20	0·94

4.3 Link Sampling

As an alternative to double sampling, a procedure (link sampling) was introduced by Harishchandra and Srivenkataramana (1982) (briefly HS (1982)) based on the idea that neighbouring lots in the production sequence often have substantial similarities, if not exactly the same quality. We therefore take a sequence of samples from a sequence of lots. Since we have to make a decision on each lot, we will examine the neighbouring lots anyway. For example, if we want to reach decisions on 100 lots, we may have to take up to 200 samples using a double sampling scheme but only up to 100 (or maybe 102) samples with the HS procedure. This can result in great savings on

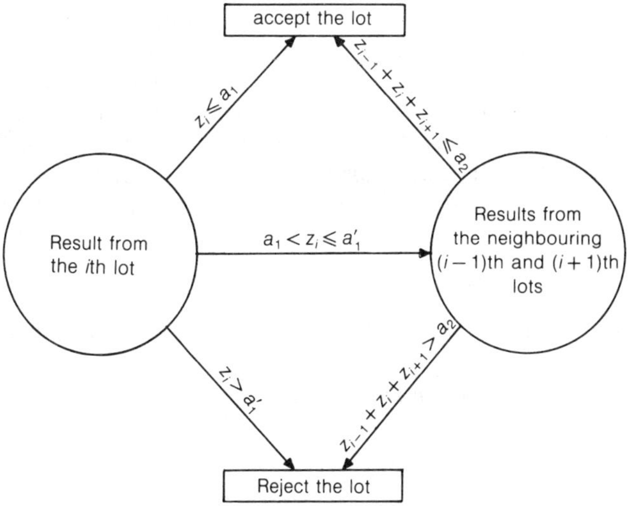

Figure 4.2 *Flow chart for link sampling*

sampling cost, although the probabilities of acceptance are altered. Here is a formal description of the HS procedure.

Suppose random samples of size n are taken without replacement from corresponding lots of size N each. Let p (p') be the probability of classifying an item as NC correctly (incorrectly), and Y_j and Z_j be the numbers of actual NC items and judged NC items, respectively, in the jth sample, from the jth lot, which has D_j actual NC items. The similarity of neighbouring lots is reflected by the similarity of values of D_{i-1}, D_i and D_{i+1}. The HS procedure uses the values of Z_{i-1} and Z_{i+1}, together with Z_i, to determine whether to accept the middle (the ith) lot. Specifically (as illustrated in Fig. 4.2):

1. if $Z_i \leqslant a_1$ accept the lot;
2. if $Z_i > a'_1$ reject the lot;
3. if $a_1 < Z_i \leqslant a'_1$ and $Z_{i-1} + Z_i + Z_{i+1} \leqslant a_2$ accept the lot;
4. if $a_1 < Z_i \leqslant a'_1$ and $Z_{i-1} + Z_i + Z_{i+1} > a_2$ reject the lot.

The numbers a_1, a'_1, and a_2 are integers chosen to give some desired probabilities of acceptance for specified proportions (D_i/N) of NC items. Typically, though not necessarily, $a_2 = a'_1$. Corresponding to $N \to \infty$, or the binomial sampling case, if the NC proportions are the same for all lots, then the acceptance probabilities for link sampling are exactly the same as they would be for ordinary double sampling with the same a_1, a'_1, and a_2, when a second sample of size $2n$ is taken from the ith lot. When N is finite, as we assumed here, we could not reach a similar conclusion even if $D_{i-1} = D_i = D_{i+1}$. The reason is that the distribution of the convolution $\text{Hypg}(n, D, N) * \text{Hypg}(n, D, N)$ is not the same as that of $\text{Hypg}(2n; D; N)$, or of $\text{Hypg}(2n; 2D; 2N)$, unlike the binomial case (when $N \to \infty, (D/N \to \omega)$) where

$$\text{Bin}(n, \omega) * \text{Bin}(n, \omega) \equiv \text{Bin}(2n, \omega).$$

The variables Z_{i-1}, Z_i and Z_{i+1} are mutually independent, and

$$Z_j \sim (\text{Bin}(Y_j, p) * \text{Bin}(n - Y_j, p')) \bigwedge_{Y_j} \text{Hypg}(n, D_j, N), \tag{4.2}$$

or, explicitly,

$$\begin{aligned} P(z_j | D_j) &\equiv \Pr[Z_j = z_j | D_j] \\ &= \binom{N}{n}^{-1} \sum_y \binom{D_j}{y} \binom{N - D_j}{n - y} \sum_{u=0}^{y} \binom{y}{u} \binom{n-y}{z_j - u} \\ &\quad \times p^u p'^{z_j - u} (1-p)^{y-u} (1-p')^{n-y-z_j+u}. \end{aligned} \tag{4.3}$$

The probability of acceptance for the ith lot is:

$$\begin{aligned} &\Pr[Z_i \leqslant a_1 | D_i] + \Pr[(a_1 < Z_i \leqslant a'_1) \\ &\qquad \cap (Z_{i-1} + Z_i + Z_{i+1} \leqslant a_2) | D_{i-1}, D_i, D_{i+1}] \\ &= \Pr[Z_i \leqslant a_1 | D_i] + \sum_{z=a_1+1}^{a'_1} \Pr[Z_i = z | D_i] \\ &\qquad \times \Pr[Z_{i-1} + Z_{i+1} \leqslant a_2 - z | D_{i-1}, D_{i+1}] \\ &= \sum_{z_i=0}^{a_1} P(z_i | D_i) + \sum_{z_i=a_1+1}^{a'_1} P(z_i | D_i) \\ &\qquad \times \sum_{z_{i-1}=0}^{a_2 - z_i} \sum_{z_{i+1}=0}^{a_2 - z_i - z_{i-1}} P(z_{i-1} | D_{i-1}) P(Z_{i+1} | D_{i+1}), \end{aligned} \tag{4.4}$$

where $P(z_j | D_j)$ is defined in (4.3).

The expected number of inspected items from the ith lot for the

HS procedure is

$$n\{1 + \Pr[a_1 < Z_i \leqslant a_1']\},$$

compared with

$$n\{1 + 2\Pr[a_1 < Z_i \leqslant a_1']\}$$

for double sampling with $n_1 = n$, $n_2 = 2n$.

Table 4 and Figs 6–7c of Johnson *et al.* (1986) illustrate values of the acceptance probability for link sampling with the same p and p' as in Table 3 and various values of D_{i-1}, D_i and D_{i+1}. If $D_{i-1} = D_i = D_{i+1} = D$, the results of link sampling are very nearly the same as those for double sampling with $n_1 = n, n_2 = 2n$. Such differences as do exist are, of course, more marked the smaller the lot size (N).

Further details can be found in Osanaiye and Alebiosu (1988). (These authors use a Poisson approximation to the binomial distribution in their calculations.)

4.4 Partial link sampling

Sometimes, when $a_1 < Z_i \leqslant a_1'$, waiting for the result of inspection of the next ($(i+1)$th) sample before reaching the decision for the present (ith) sample is not desirable. A natural idea is to take a sample from the present lot again, so that no waiting is needed. HS (1982) propose a partial linked sampling procedure, in which a second sample is taken from the ith lot instead of waiting for the $(i+1)$th. Suppose Z_i' is the number of items declared NC in the second sample of size n from the ith lot. This procedure can be obtained by just replacing Z_{i+1} by Z_i' in steps 3 and 4 of the link sampling procedure in section 4.3. The acceptance probability is

$$\Pr[Z_i \leqslant a_1] + \Pr[(a_1 < Z_i \leqslant a_1') \cap (Z_{i-1} + Z_i + Z_i' \leqslant a_2)]. \quad (4.5)$$

Now although Z_{i-1} is independent of (Z_i, Z_i'), Z_i and Z_i' are not. With Y_i' denoting the actual number of NC items in the second sample from the ith lot, the joint distribution of Z_i and Z_i' is

$$\begin{pmatrix} Z_i \\ Z_i' \end{pmatrix} \sim \begin{pmatrix} \text{Bin}(Y_i, p) * \text{Bin}(n - Y_i, p') \\ \text{Bin}(Y_i', p) * \text{Bin}(n - Y_i', p') \end{pmatrix}_{Y_i, Y_i'} \bigwedge \text{Mult Hypg}(n, n; D_i; N). \quad (4.6)$$

Here Mult Hypg$(n, n; D_i; N)$ denotes a multivariate hypergeometric distribution with

$$\Pr[(Y_i = y_i) \cap (Y_{i'} = y_{i'})] = \binom{n}{y_i}\binom{n}{y_i'}\binom{N-2n}{D-y_i-y_{i'}} \bigg/ \binom{N}{D_i}, \tag{4.7}$$

where $0 \leqslant y_i, y_{i'} \leqslant n; D_i - N + 2n \leqslant y_i + y_{i'} + D_{i'}$.

The expected number of inspected items from the ith lot is

$$n\{1 + \Pr[z_1 < Z_i \leqslant a_1']\} = \tfrac{1}{2}[n + n\{1 + 2\Pr[a_1 < Z_i \leqslant a_1']\}],$$

compared with

$$n\{1 + 2\Pr[a_1 < Z_i \leqslant a_1']\}$$

for double sampling with $n_1 = n, n_2 = 2n$.

Table 5 of Johnson *et al.* (1986) gives values of the probabilities of acceptance for partial link sampling, which are comparable with those for link sampling (cf. Table 4).

Conclusion

From the analytic expression for acceptance probability, it is hard to see the direction of the influence caused by the effects of varying parameters such as p, p', N, n and D. However the computation results provide quite a clear picture.

First, for a wide range of both the double sampling and link sampling procedures, the probabilities of acceptance are insensitive to p as low as 0.95 while it is sensitive to p' as small as 0.01. Second, for double sampling schemes, when the sample size is large, the effects of changes in p and p' are more significant, but the lot size has little effect on the probabilities of acceptance for given D/N. Finally, for link sampling procedures, the probabilities of acceptance are naturally sensitive to differences between D_i and the average of D_{i-1} and D_{i+1}. All of the above features of these schemes are intuitively reasonable, but they serve as reminders of the delicate relationships among the parameters and sensitivity to values of the error probabilities.

4.5 Continuous sampling plans (CSP)

When it is possible to inspect items sequentially, in order as they are produced, 'continuous' sampling plans may be used. A particular set of such plans, developed from the work of Dodge (1943, 1970) and Dodge and Torrey (1951) has been widely used. We first give a brief description of the more common of these plans (CSP-1, CSP-2, CSP-3) and then evaluate the effects of inspection errors on their properties.

All three plans commence with 100% inspection, which continues until i successive items are deemed to be C. (i is known as the **clearing interval**.) Then sampling inspection, applied to a fraction f (the **sampling fraction**) of the output commences. This continues until an NC decision occurs.

The next step varies among the three plans.

In CSP-1 (the basic plan) there is immediate reversion to 100% inspection, and a new cycle begins.

In CSP-2, reversion is delayed to give the production process a chance to rehabilitate itself. Sampling inspection continues. If a further NC decision occurs before $(k + 1)$ more items are inspected, there is immediate reversion to 100% inspection. If, on the other hand, no such decision occurs among the first k additional items inspected, the status of the plan reverts to the beginning of the sampling inspection (so that this is continued indefinitely until an NC decision occurs, and so on).

The third sampling plan, CSP-3, is a further modification of CSP-2. A four-item sequence of 100% inspection is inserted before beginning the additional k items under sampling inspection. If any of the four items gives an NC decision, 100% inspection is reinstated immediately after the four-item sequence, and a new cycle starts. Otherwise, we proceed as in CSP-2.

It can be seen that, while CSP-2 is more lenient than CSP-1, CSP-3 is less lenient than CSP-2 in that $(k + 4)$ items have to be declared C for the original sampling inspection status to be regained.

We will investigate

1. the expected length of a cycle between successive initiations of unlimited 100% inspection;
2. the expected proportion of items inspected; and
3. the average outgoing quality (AOQ) as measured by the expected proportion of truly NC items among those accepted (on the assumption that all items deemed to be NC are replaced by items known to be C).

It is clear that appropriate formulas for 1 and 2 can be derived from corresponding formulas for error-free inspection by replacing ω (the proportion of NC in the population) by

$$\bar{p} = \omega p + (1 - \omega)p' \qquad \text{(cf. equation (2.14))}$$

(the probability that a randomly chosen item is declared to be NC on inspection – this is effective fraction defective ($\tilde{\omega}$) of Chapter 1).

Stephens (1979) gives results for error-free inspection obtained by using Markov chain theory, Case *et al.* (1973) deal with CSP-1, allowing for inspection error, and Chakraborty and Rathie (1990) obtain results for CSP-1, -2 and -3, also allowing for inspection error, using a stochastic process technique known as GERT. (See also Chakraborty and Bapaye (1989); Chakraborty and Mondal (1985).) We will use direct methods, of the kind used earlier in this book.

1. Each cycle is made up of

   (a) an initial component of 100% inspection followed by
   (b) a 'sampling inspection' component (including, in the case of CSP-3, one or more four-item 100% inspection sequences).

   For all three sampling plans, the expected length of (a) (in items produced) is

$$\begin{aligned} E_1 &= (1-\bar{p})^{-1} + (1-\bar{p})^{-2} + \cdots + (1-\bar{p})^{-i} \\ &= \{\bar{p}(1-\bar{p})^i\}^{-1}\{1-(1-\bar{p})^i\} \end{aligned} \tag{4.8}$$

   (Bizley (1962)).

   For CSP-1 the expected number of items inspected to encounter an NC decision is $1/\bar{p}$, corresponding to

$$E_{f1} = (f\bar{p})^{-1} \text{ production items, in (b),}$$

   For CSP-1, therefore, the expected length of a cycle is

$$E_1 + E_{f1} = \bar{p}^{-1}\{(1-\bar{p})^{-i} - 1 + f^{-1}\}. \tag{4.9}$$

   For CSP-2 and CSP-3 the sampling inspection component is made up from possible subcycles, each of expected length (in production items)

$$(f\bar{p})^{-1} + E'_{fj}$$

   where E'_{fj} is the expected length of the additional inspections interposed between occurrence of the first NC item in sampling inspection and reinstatement of original sampling inspection status or return to 100% inspection, as the case may be.

For CSP-2 the probability of reinstatement to a further subcycle is $(1-\bar{p})^k$; for CSP-3 it is $(1-\bar{p})^{k+4}$, so the expected length (in production items) of the whole sampling inspection component is

$$E_{f2} = \{(f\bar{p})^{-1} + E'_{f2}\}[1 + (1-\bar{p})^k + (1-\bar{p})^{2k} + \cdots]$$

$$= \{(f\bar{p})^{-1} + E'_{f2}\}\{1-(1-\bar{p})^k\}^{-1} \qquad \text{for CSP-2} \qquad (4.10a)$$

and

$$E_{f3} = \{(f\bar{p})^{-1} + E'_{f3}\}\{1-(1-\bar{p})^{k+4}\}^{-1} \qquad \text{for CSP-3.} \qquad (4.10b)$$

In both CSP-2 and -3, the expected number of items inspected in the sequence of up to k additional items (given that this is reached for CSP-3) can be computed in the following way. It is the expected number $(\bar{p}^{-1})$ of inspections needed in an *unlimited* series to obtain the first NC decision, less the expected number in excess of k in such a sequence, which is $(1-\bar{p})^k\bar{p}^{-1}$.

For CSP-2, therefore

$$E'_{f2} = \{1-(1-\bar{p})^k\}/(f\bar{p}) \qquad (4.11a)$$

while for CSP-3

$$E'_{f3} = 4 + (1-\bar{p})^4\{1-(1-\bar{p})^k\}/(f\bar{p}) \qquad (4.11b)$$

(since the probability of reaching the k additional inspections is $(1-\bar{p})^4$).

Combining equations (4.10a, b) and (4.11a, b) we obtain the expected length of cycles for CSP-j as

$$E_1 + E_{fj}, \qquad (j = 2, 3),$$

with

$$E_{f2} = (f\bar{p})^{-1}\{2-(1-\bar{p})^k\}\{1-(1-\bar{p})^k\}^{-1} \qquad \text{for CSP-2} \qquad (4.12a)$$

and

$$E_{f3} = (f\bar{p})^{-1}[1 + 4f\bar{p} + (1-\bar{p})^4\{1-(1-\bar{p})^k\}]\{1-(1-\bar{p})^{k+4}\}^{-1} \qquad \text{for CSP-3.} \qquad (4.12b)$$

Note that in E_{f3} there is a contribution of

$$4\{1-(1-\bar{p})^{k+4}\}^{-1}$$

from 100% inspection within the 'sampling inspection' component.

2. The proportion of items inspected is

$$(E_1 + fE_{fj})/(E_1 + E_{fj}) \qquad \text{for CSP-}j\ (j = 1, 2)$$

and

$$\frac{1}{E_1 + E_{f3}}\left[E_1 + \frac{4}{1-(1-\bar{p})^{k+4}} + f\left\{E_{f3} - \frac{1}{1-(1-\bar{p})^{k+4}}\right\}\right]$$

$$= \frac{1}{E_1 + E_{f3}}\left\{E_1 + \frac{4(1-f)}{1-(1-\bar{p})^{k+4}} + fE_{f3}\right\} \qquad \text{for CSP-3.}$$

3. The average outgoing quality does depend explicitly on quantities other tna $\bar{p}$. Under our assumption that all items declared NC are replaced by truly C items, the accepted items which are, in reality, NC are composed of a proportion ω of those items which are not inspected plus a proportion $\Pr[\text{NC}|\text{declared C}] = \omega(1-p)/(1-\bar{p})$, (cf. equation (5.16); an application of Bayes' theorem) of those inspected and declared C.

   Hence

$$\text{AOQ} = \begin{cases} \dfrac{\omega(1-p)}{1-\bar{p}}(E_1 - i) + \omega(1-f)E_{fj} & \text{for CSP-}j\ (j=1,2) \\ \dfrac{\omega(1-p)}{1-\bar{p}}\left\{E_1 - i + \dfrac{4}{1-(1-\bar{p})^{k+4}}\right\} & \\ \quad + \omega(1-f)\left\{E_{f3} - \dfrac{4}{1-(1-\bar{p})^{k+4}}\right\} & \text{for CSP-3.} \end{cases}$$

Clearly, CSP-3 can be varied by replacing the four-items sequence by one involving d($\neq 4$) items. The reader is encouraged to investigate numerically the effects of errors on these sampling plans. (We believe the derivations presented here have not appeared previously in the literature.

CHAPTER 5

Multitype nonconformities

5.1 Introduction

In previous chapters, an item can be either C or NC, and there is only one type of nonconformity. But in practice each item may have more than one kind of nonconformity at the same time. For example, a metal material may have (or not have) surface irregularities and/or internal flaws. This example represents the simplest case, where there exist two types of nonconformity, which we will discuss first. We will also study more general cases where there are more than two types of nonconformity. An interesting example of multiple defects is described by Bonett (1988). This relates to new automobiles at the moment of leaving the production line. Defects of eight different types are identified: (1) torn carpet; (2) inoperative right brake light; (3) improperly installed glove box latch; (4) cracked emergency brake handle, (5) missing bolt on front of passenger seat; (6) bubble in roof; (7) loose right side molding; and (8) inoperative left real speaker. As before, we are interested in the cases where inspection is not perfect.

In the next two sections, we detect only one particular type of nonconformity among two or more existing NC types, and we will discuss the distributions and moments of related variables.

In the final section, we are interested in detecting not only one type of nonconformity among others but patterns of nonconformities. (A **pattern** is a combination of different NC types.) For each item, there are only two possibilities in regard to any single NC type: presence or absence. If there are k NC types, there are, in all 2^k patterns which an item may have.

5.2 Inspection for one NC type among two

Notation

For simplicity, the phrase 'type (i)' will stand for 'type (i) nonconformity' below. Suppose that a lot has N items, and there exist two types of nonconformity, types (1) and (2), and in the lot, there are D_{gh} items with g type (1) and h type (2), $(g, h = 0, 1)$. For example, D_{00} is the number of items with no nonconformities; D_{01} is the number of items having only type (2); etc. Obviously $D_{00} + D_{01} + D_{10} + D_{11} = N$. We define the vector $\mathbf{D} \equiv (D_{00}, D_{01}, D_{10}, D_{11})$, which has $2^2 = 4$ elements, each element corresponding to an NC pattern.

Let n be the size of a sample, randomly selected from the lot without replacement. Suppose that inspection is only for type (1) nonconformity. Let Z_1 be the number of items which are declared to have type (1) after inspection. Among these Z_1 items, let Z_1^* and Z_2^* denote the numbers of items actually having type (1) and type (2), respectively. In other words, Z_1 items are judged to belong to pattern (1, 1) or (1, 0), and among those Z_1 items, there are Z_1^* (resp. Z_2^*) items actually having pattern (1, 1) or (1, 0) (resp. (1, 1) or (0, 1)). We also suppose that the probability of correct (false) detection of type (i) nonconformity is p_i (p_i'). In the sample, the counterpart of $\mathbf{D}$ in the lot is $\mathbf{Y} \equiv (Y_{00}, Y_{01}, Y_{10}, Y_{11})$; $\mathbf{y} \equiv (y_{00}, y_{01}, y_{10}, y_{11})$ is a realized value of $\mathbf{Y}$. Of course,

$$Y_{00} + Y_{01} + Y_{10} + Y_{11} = y_{00} + y_{01} + y_{10} + y_{11} = n.$$

Let Y_1 be the number of items which really have type (1), i.e. $Y_1 = Y_{10} + Y_{11}$, and similarly let $y_1 = y_{10} + y_{11}$.

Distributions and moments

It is well known that

$$Y_1 = Y_{10} + Y_{11} \sim \text{Hypg}(n; D_1^*, N),$$

where $D_1^* \equiv D_{10} + D_{11}$. Let W_{gh} be the number of items having pattern (gh) that are judged to possess type (1). Conditional on $\mathbf{Y} = \mathbf{y}$, the W_{gh}'s are mutually independent and

$$W_{gh} \sim \text{Bin}(y_{gh}, gp_1 + (1 - g)p_1'),$$

for $g, h = 0, 1$. It is also easy to see that given $\mathbf{Y} = \mathbf{y}$, Z_1 is distributed as the sum of two independent binomial variables. One of them is the number of correctly classified items which really have type (1), which is $W_1 \equiv W_{10} + W_{11}$, and conditionally

$$W_1 | \mathbf{Y} = \mathbf{y} \sim \mathrm{Bin}(y_{10} + y_{11}, p_1) = \mathrm{Bin}(y_1, p_1).$$

The other one is the number of incorrectly classified items, not really possessing type (1), which is $W_0 \equiv W_{01} + W_{00}$, and conditionally

$$W_0 | \mathbf{Y} = \mathbf{y} \sim \mathrm{Bin}(y_{00} + y_{01}, p'_1) = \mathrm{Bin}(n - y_1, p'_1).$$

Note that the condition $\mathbf{Y} = \mathbf{y}$ is equivalent to $Y_1 = y_1$ for the distributions of W_1, W_0 and Z_1. We therefore have

$$Z_1 = W_0 + W_1 = W_{00} + W_{01} + W_{10} + W_{11}, \tag{5.1}$$

and

$$Z_1 | Y_1 = y_1 \sim \mathrm{Bin}(n - y_1, p'_1) * \mathrm{Bin}(y_1, p_1). \tag{5.2}$$

We also have the overall distribution of Z_1, namely

$$Z_1 \sim [\mathrm{Bin}(Y_1, p_1) * \mathrm{Bin}(n - Y_1, p'_1)] \underset{Y_1}{\wedge} \mathrm{Hypg}(n; D_1^*, N).$$

Our interest here is not in Z_1 itself, because that corresponds to the case of a single type of nonconformity, already discussed in earlier chapters. We want to know the properties of certain other variables and their interrelations. We have already introduced

$$Z_1^* = W_{10} + W_{11} \tag{5.3}$$

and

$$Z_2^* = W_{01} + W_{11}. \tag{5.4}$$

Distributions of Z_1^* and Z_2^* are those of convolutions of the corresponding W_{gh}'s.

The distribution of Z_1^* is the hypergeometric–binomial;

$$Z_1^* \sim \mathrm{Bin}(Y_1, p_1) \underset{Y_1}{\wedge} \mathrm{Hypg}(n; D_1^*, N). \tag{5.5}$$

The joint probability that the sample will contain y_{gh} items with g type (1) and h type (2) nonconformities ((gh) pattern; $g, h = 0, 1$) is

$$\Pr[\mathbf{Y} = \mathbf{y}] = \Pr\left[\prod_{g=0}^{1} \prod_{h=0}^{1} (Y_{gh} = y_{gh})\right] = \left[\prod_{g=0}^{1} \prod_{h=0}^{1} \binom{D_{gh}}{y_{gh}}\right] \Big/ \binom{N}{n}, \tag{5.6}$$

a multivariate hypergeometric $(n; \mathbf{D}, N)$ distribution. The $\boldsymbol{\alpha} = (\alpha_{00}, \alpha_{01}, \alpha_{10}, \alpha_{11})$th joint factorial moment of $\mathbf{Y}$ is

$$\mu_{(\boldsymbol{\alpha})}(\mathbf{Y}) \equiv E\left[\prod_{g=0}^{1}\prod_{h=0}^{1} Y_{gh}^{(\alpha_{gh})}\right] = n^{(b)}\left[\prod_{g=0}^{1}\prod_{h=0}^{1} D_{gh}^{(\alpha_{gh})}\right]\Big/ N^{(b)}, \tag{5.7}$$

where $b = \alpha_{00} + \alpha_{01} + \alpha_{10} + \alpha_{11}$. From (5.1)–(5.4) we easily obtain the following results.

$$E[Z_1 | \mathbf{y}] = (y_{10} + y_{11})p_1 + (y_{00} + y_{01})p_1';$$
$$\mathrm{Var}(Z_1 | \mathbf{y}) = (y_{10} + y_{11})p_1(1 - p_1) + (y_{00} + y_{01})p_1'(1 - p_1'), \tag{5.8}$$

$$E[Z_2^* | \mathbf{y}] = y_{11}p_1 + y_{01}p_1';$$
$$\mathrm{Var}(Z_2^* | \mathbf{y}) = y_{11}p_1(1 - p_1) + y_{01}p_1'(1 - p_1'), \tag{5.9}$$

$$\mathrm{cov}(Z_1, Z_2^* | \mathbf{y}) = \mathrm{var}(W_{01} + W_{11} | \mathbf{y}) = \mathrm{var}(Z_2^* | \mathbf{y}), \tag{5.10}$$

$$\mathrm{corr}(Z_1, Z_2^* | \mathbf{y}) = \{\mathrm{var}(Z_2^* | \mathbf{y})/\mathrm{var}(Z_1 | \mathbf{y})\}^{1/2}, \tag{5.11}$$

$$\mathrm{cov}(Z_1^*, Z_2^* | \mathbf{y}) = \mathrm{var}(W_{11} | \mathbf{y}) = y_{11}p_1(1 - p_1), \tag{5.12}$$

$$\mathrm{corr}(Z_1^*, Z_2^* | \mathbf{y}) = \left[\frac{y_{11}^2 p_1(1 - p_1)}{(y_{10} + y_{11})\{y_{11}p_1(1 - p_1) + y_{01}p_1'(1 - p_1')\}}\right]^{1/2}. \tag{5.13}$$

(Recall $Z_1^* | \mathbf{y} \sim \mathrm{Bin}(y_{10} + y_{11}, p_1)$.)
The distribution of Z_2^* is the multivariate hypergeometric–convoluted binomial;

$$Z_2^* \sim \{\mathrm{Bin}(Y_{11}, p_1) * \mathrm{Bin}(Y_{01}, p_1')\} \underset{Y_{11}, Y_{01}}{\wedge} \mathrm{Mult\,Hypg}(n; D_{11}, D_{01}; N). \tag{5.14}$$

Using the methods introduced in Chapter 2, moments can be obtained.

For an item chosen at random from the lot,

$$\begin{aligned}&\Pr[\text{classified as having type (1)}]\\ &= N^{-1}\{(D_{01} + D_{00})p_1' + (D_{10} + D_{11})p_1\}\\ &= N^{-1}(D_0 p_1' + D_1 p_1) = \bar{p}_1\end{aligned} \tag{5.15}$$

say, $(D_1 \equiv D_{10} + D_{11}$ and $D_0 \equiv N - D_1)$, and

$$\Pr[\text{actually having type (2)} | \text{classified as having type (1)}]$$
$$= \frac{N^{-1}(D_{01}p_1' + D_{11}p_1)}{\bar{p}_1} = \frac{\bar{p}_{2|1}}{\bar{p}_1}, \tag{5.16}$$

say. Hence

$$E[Z_2^*|Z_1] = Z_1 \frac{\bar{p}_{2|1}}{\bar{p}_1}, \tag{5.17}$$

and from (3.14) (or (5.15) and (5.16) above)

$$E[Z_2^*] = n\bar{p}_{2|1}. \tag{5.18}$$

Conditioned on $\mathbf{Y} = \mathbf{y}$, formula (5.17) is also true, and therefore

$$\text{cov}(Z_1, Z_2^*) = \frac{\bar{p}_{2|1}}{\bar{p}_1} E[Z_1^2] - n^2 \bar{p}\bar{p}_{2|1}, \tag{5.19}$$

and from (3.15) and (3.14)

$$\text{cov}(Z_1, Z_2^*) = \frac{n\bar{p}_{2|1}}{\bar{p}_1} \left\{ \bar{p}_1(1 - \bar{p}_1) - \frac{(n-1)D_1}{(N-1)N} \left(1 - \frac{D_1}{N}\right)(p_1 - p_1')^2 \right\}. \tag{5.20}$$

Since $E[Z_2^*|Z_1]$ is increasing with Z_1 by (5.17), $\text{cov}(Z_1, Z_2^*) > 0$.

5.3 Inspection for one NC type in multitype nonconformity case

As in the preceding section, inspection is for type (1) nonconformity only, but there are $m(\geqslant 2)$ other types of nonconformity, (2),..., $(m + 1)$, as well as type (1). Let $Z_j^*(j = 2, \ldots, m + 1)$ be the number of items having type (j) that are classified as having type (1) nonconformity. Extension from $m = 1$ to $m \geqslant 2$ is direct. The novel issue is to find the joint distribution of $Z_2^*, \ldots, Z_{m+1}^*$. Using the '0-1' notation system similar to that in section 5.2, we have, for example, (corresponding to (5.16)) for an item randomly chosen from the lot:

$$\begin{aligned}&\Pr[\text{having NC types (2) and (3),}\\ &\quad \text{but not } (4), \ldots, (m+1)|\\ &\quad \text{classified as being NC (1)}]\\ &= \frac{D_{0110\cdots0}p_1' + D_{1110\cdots0}p_1}{D_{0\cdots}p_1' + D_{1\cdots}p_1},\end{aligned}$$

where

$$D_{g\cdots} = \sum_{a_2=0}^{1} \cdots \sum_{a_{m+1}=0}^{1} D_{ga_2\cdots a_{m+1}}, \qquad (g = 0, 1) \tag{5.21}$$

is the total number of items with g type (1) nonconformity in the lot. (The summation is over all D's with g as first subscript.)

Distributions and moments

For simplicity, we shall consider only the case $m = 2$, i.e. there are three NC types. Our aim is to find an expression for $\text{cov}(Z_2^*, Z_3^*)$. We will first find the joint distribution of Z_2^* and Z_3^*, and then the joint factorial moment $E[Z_2^{*(s_2)} Z_3^{*(s_3)}]$, and finally the covariance. At each step, we calculate the conditional moment (or covariance) first, and then the overall moment. The notation is similar to that in the previous section. We will then give an approximation for the covariance.

Let $\mathbf{Y} \equiv (Y_{111}, Y_{011}, Y_{101}, Y_{110}, Y_{100}, Y_{010}, Y_{001}, Y_{000})$, and let W_{ghi}'s $(g, h, i = 0, 1)$ be eight independent binomial variables with parameters $(y_{ghi}, gp_1 + (1-g)p_1')$. The meanings of $\mathbf{Y}$, $\mathbf{y}$, and the W_{ghi}'s are analogous to those in section 5.2. Analogously to (5.3) and (5.4), conditional on $\mathbf{Y} = \mathbf{y}$,

$$Z_2^* = W_{111} + W_{011} + W_{110} + W_{010};$$
$$Z_3^* = W_{111} + W_{011} + W_{101} + W_{001}. \tag{5.22}$$

Symbolically,

$$\begin{Bmatrix} Z_2^* \\ Z_3^* \end{Bmatrix} \sim \text{Bin}(Y_{111}, p_1) * \text{Bin}(Y_{011}, p_1') * \begin{Bmatrix} \text{Bin}(Y_{110}, p_1) * \text{Bin}(Y_{010}, p_1') \\ \text{Bin}(Y_{101}, p_1) * \text{Bin}(Y_{001}, p_1') \end{Bmatrix}$$
$$\bigwedge_{\mathbf{Y}} \text{Mult Hypg}(n; D, N), \tag{5.23}$$

where the first two binomial distributions arise from the common terms, W_{111} and W_{011}, and the next two pairs are from the terms which Z_2^* and Z_3^* do not share.

To obtain a general expression of the joint factorial moment $E[Z_2^{*(s_2)} Z_3^{*(s_3)}]$, we first note the factorial expression:

$$\begin{aligned}
Z_2^{*(s_2)} Z_3^{*(s_3)} &= \left\{ \sum_i^{(4)} \binom{s_2}{\mathbf{i}} W_{111}^{(i_1)} W_{011}^{(i_2)} W_{110}^{(i_3)} W_{010}^{(i_4)} \right\} \\
&\quad \times \left\{ \sum_j^{(4)} \binom{s_3}{\mathbf{j}} W_{111}^{(j_1)} W_{011}^{(j_2)} W_{101}^{(j_3)} W_{001}^{(j_4)} \right\} \\
&= \sum_i^{(4)} \sum_j^{(4)} \binom{s_2}{\mathbf{i}} \binom{s_3}{\mathbf{j}} \left\{ \sum_{g=0}^{j_1} \binom{j_1}{g} i_1^{(j_1-g)} W_{111}^{(i_1+g)} \right\} \\
&\quad \times \left\{ \sum_{h=0}^{j_2} \binom{j_2}{h} i_2^{(j_2-h)} W_{011}^{(i_2+h)} \right\} W_{110}^{(i_3)} W_{010}^{(i_4)} W_{101}^{(j_3)} W_{001}^{(j_4)},
\end{aligned} \tag{5.24}$$

where

$$\sum_{i}^{(4)} = \sum_{i_1}\sum_{i_2}\sum_{i_3}\sum_{i_4} \quad \text{with constraint} \quad \sum_{l=1}^{4} i_l = s_2;$$

$$\sum_{j}^{(4)} = \sum_{j_1}\sum_{j_2}\sum_{j_3}\sum_{j_4} \quad \text{with constraint} \quad \sum_{l=1}^{4} j_l = s_3;$$

This is derived from the identity

$$a^{(\alpha_1)}a^{(\alpha_2)} = a^{(\alpha_1)}(a-\alpha_1+\alpha_1)^{(\alpha_2)} = \sum_{u=0}^{\alpha_2}\binom{\alpha_2}{u}\alpha_1^{(\alpha_2-u)}a^{(\alpha_1+u)}.$$

Since the W_{ghi}'s are mutually independent, replacing $W_{ghi}^{(\alpha)}$ in (5.24) by $Y_{ghi}^{(\alpha)}\tilde{p}_g$, where $\tilde{p}_g = p'_1$ if $g = 1$ and p_1 if $g = 0$, gives an expression for $E[Z_2^{*(s_2)}Z_3^{*(s_3)}|\mathbf{Y}=\mathbf{y}]$, and, further, taking expectation with respect to $\mathbf{Y}$ leads to

$$\begin{aligned} E[Z_2^{*(s_2)}Z_3^{*(s_3)}] = &\sum_{i}^{(4)}\sum_{j}^{(4)}\sum_{g=0}^{j_1}\sum_{h=0}^{j_2}\binom{s_2}{\mathbf{i}}\binom{s_3}{\mathbf{j}}\binom{j_1}{g}\binom{j_2}{h} \\ &\times p_1^{g+i_1+i_3+j_3}p_1'^{h+i_2+i_4+j_4} \\ &\times i_1^{(j_1-g)}i_2^{(j_2-h)}\frac{n^{(s_2+s_3+g+h)}}{N^{(s_2+s_3+g+h)}} \\ &\times D_{111}^{(i_1+g)}D_{011}^{(i_2+h)}D_{110}^{(i_3)}D_{101}^{(i_4)}D_{010}^{(j_3)}D_{001}^{(j_4)}. \end{aligned} \tag{5.25}$$

For cases with $m > 2$, the same expectation could be reached, with D_{ghi} replaced by $D_{ghi\cdots}$; a summation of D's over all patterns with ghi as first three subscripts. Joint factorial moments of three or more Z_i^*'s $(i > 1)$ can be similarly obtained, though the formulas rapidly become more cumbersome. However, they are in a form suitable as a basis for computer programs.

General formulas are not necessarily simple to use; for lower order moments, direct calculation is often easier.

For calculating $\text{cov}(Z_2^*, Z_3^*)$, from (5.22), we have

$$\begin{aligned} \text{cov}(Z_2^*, Z_3^*|\mathbf{y}) &= \text{Var}(W_{111}+W_{011}|\mathbf{y}) \\ &= y_{111}p_1(1-p_1) + y_{011}p'_1(1-p'_1), \end{aligned} \tag{5.26}$$

whence

$$\begin{aligned} E[Z_2^*Z_3^*|\mathbf{y}] &= \text{cov}(Z_2^*, Z_3^*|\mathbf{y}) + E[Z_2^*|\mathbf{y}]E[Z_3^*|\mathbf{y}] \\ &= y_{111}p_1(1-p_1) + y_{011}p'_1(1-p'_1) \\ &\quad + \{(y_{111}+y_{110})p_1 + (y_{011}+y_{010})p'_1\} \\ &\quad \times \{(y_{111}+y_{101})p_1 + (y_{011}+y_{001})p'_1\}, \end{aligned}$$

and, recalling that the **Y**'s have a multivariate hypergeometric distribution,

$$\begin{aligned}\operatorname{cov}(Z_2^*, Z_3^*) &= E[Z_2^* Z_3^*] - E[Z_2^*]E[Z_3^*] \\ &= \frac{n}{N}\{D_{111}p_1(1-p_1) + D_{011}p_1'(1-p_1')\} \\ &\quad + \left\{\frac{n}{N} - \frac{n(n-1)}{N(N-1)}\right\}(D_{111}p_1^2 + D_{011}p_2'^2) \\ &\quad + \left\{\frac{n(n-1)}{N(N-1)} - \frac{n^2}{N^2}\right\}\{(D_{111}+D_{011})p_1 \\ &\quad + (D_{011}+D_{010})p_1'\}\{(D_{111}+D_{101})p_1 \\ &\quad + (D_{011}+D_{001})p_1'\} \\ &= \frac{n}{N}\left\{D_{111}p_1\left(1 - \frac{n-1}{N-1}p_1\right)\right. \\ &\quad \left. + D_{011}p_1'\left(1 - \frac{n-1}{N-1}p_1'\right)\right\} - \frac{n(N-n)}{N(N-1)} \\ &\quad \times \{(D_{111}+D_{110})p_1 + (D_{011}+D_{010})p_1'\} \\ &\quad \times \{(D_{111}+D_{101})p_1 + (D_{011}+D_{001})p_1'\}, \end{aligned} \tag{5.27}$$

which can either be positive or negative. While both Z_2^* and Z_3^* are positively correlated with Z_1, they may be negatively correlated in the population, corresponding to values of D_{g10} and D_{g01} which are large, compared to D_{g11} and D_{g00} $(g = 0, 1)$.

Approximations for large N

When N is large compared with n, we may use the approximation (omitting terms of order n/N)

$$\operatorname{cov}(Z_2^*, Z_3^*) \doteq n[\bar{p}_{1|11} - (\bar{p}_{1|11} + \bar{p}_{1|10})(\bar{p}_{1|11} + \bar{p}_{1|01})], \tag{5.28}$$

where $\bar{p}_{1|hi} = N^{-1}(D_{1hi}p_1 + D_{0hi}p_1')$.

For illustration, suppose that $p_1 = 0.90$ and $p_1' = 0.10$;

$$\begin{aligned} &D_{111}/N = 0{\cdot}01, &&D_{110}/N = 0{\cdot}1, &&D_{101}/N = 0{\cdot}15, \\ &D_{011}/N = 0{\cdot}01, &&D_{010}/N = 0{\cdot}15, &&D_{001}/N = 0{\cdot}2. \end{aligned}$$

We have

$$\bar{p}_{1|11} = 0{\cdot}010, \qquad \bar{p}_{1|10} = 0{\cdot}105, \qquad \bar{p}_{1|01} = 0{\cdot}155;$$
$$\bar{p}_{1|11} - (\bar{p}_{1|11} + \bar{p}_{1|10})(\bar{p}_{1|11} + \bar{p}_{1|01}) < 0$$

corresponding to negative $\operatorname{cov}(Z_2^*, Z_3^*)$.

If we increase D_{111}/N and D_{011}/N to 0·05, leaving the other parameters unchanged, then $\bar{p}_{1|11} = 0{\cdot}050$, while $\bar{p}_{1|10}$ and $\bar{p}_{1|01}$ remain the same, and

$$\bar{p}_{1|11} - (\bar{p}_{1|11} + \bar{p}_{1|10})(\bar{p}_{1|11} + \bar{p}_{1|01}) = 0{\cdot}050 - (0{\cdot}155 \times 0{\cdot}205) > 0$$

corresponding to positive $\operatorname{cov}(Z_2^*, Z_3^*)$.

With D_{ghi} replaced by $D_{ghi\ldots}$, one can obtain similar results for $m > 2$.

Detection of one type affected by the presence or absence of others

In practice, it is quite possible that the presence or absence of one type of nonconformity may affect the chances of detection of another type (especially in medicine). Consider the case in section 5.2 with two NC types, (1) and (2). Let us consider the influence of the presence or absence of type (2) nonconformity on the chance of detection of type (1). We introduce some new notation:

$$p_{[2]1} = \Pr[\text{correct detection of (1), in presence of (2)}];$$
$$p_{(2)1} = \Pr[\text{correct detection of (1), in absence of (2)}];$$
$$p'_{[2]1} = \Pr[\text{incorrect detection of (1), in presence of (2)}];$$
$$p'_{(2)1} = \Pr[\text{incorrect detection of (1), in absence of (2)}].$$

The relations (5.1), (5.3) and (5.4) are still unchanged, except that the parameters of the binomial distribution (conditioned on $\mathbf{Y} = \mathbf{y}$) for the W_{gh}'s should be modified. We then have

$$W_{gh} \sim \begin{cases} \operatorname{Bin}(y_{g0}, gp_{(2)1} + (1-g)p'_{(2)1}) & \text{for } h = 0 \text{ ((2) absent)}, \quad g = 1, 2, \\ \operatorname{Bin}(y_{g1}, gp_{[2]1} + (1-g)p'_{[2]1}) & \text{for } h = 1 \text{ ((2) present)}, \quad g = 1, 2. \end{cases}$$

The counterpart of (5.16) is now

$$\Pr[\text{classified as having type (1)} \mid \text{having type (2)}]$$
$$\frac{D_{01}p'_{[2]1} + D_{11}p_{[2]1}}{D_{00}p'_{(2)1} + D_{01}p'_{[2]1} + D_{10}p_{(2)1} + D_{11}p_{[2]1}} = \frac{\bar{p}^*_{2|1}}{\bar{p}^*_1}, \qquad (5.29)$$

with

$$\bar{p}_1^* = \frac{1}{N}(D_{00}p'_{(2)1} + D_{01}p'_{[2]1} + D_{10}p_{(2)1} + D_{11}p_{[2]1})$$

$$= \Pr[\text{classified as being type (1)}]$$

and

$$\bar{p}_{2|1}^* = \frac{1}{N}(D_{01}p'_{[2]1} + D_{11}p_{[2]1}).$$

Replacing $\bar{p}_1$, $\bar{p}_{2|1}$ by $\bar{p}_1^*$, $\bar{p}_{2|1}^*$, respectively, many of the formulas we have obtained previously, such as (5.17) and (5.18), are still valid.

5.4 Multitype nonconformities, the general case

Introduction and Notation

We now consider the general case of inspection for k NC types, (1),..., (k). In previous sections we only considered inspection for a single NC (type (1)). In the general case, we will classify an item as having an NC **pattern**. We introduce, for a size n sample, the notation

$$Z_{g_1,\ldots,g_k} = Z_{\mathbf{g}} \qquad (g_i = 0, 1;\ i = 1,\ldots, k) \tag{5.30}$$

as the number of items judged to have g_i type (i) NC ($i = 1,\ldots, k$) after inspection. The vector $(\mathbf{g}) = ((g_1,\ldots, g_k))$ denotes an 'NC pattern'. We shall investigate the joint distribution of the $Z_{\mathbf{g}}$'s, and also of the random variables

$$Z^*_{h_1,\ldots\, h_k(g_1,\ldots,g_k)} = Z^*_{\mathbf{h}(\mathbf{g})}, \tag{5.31}$$

which represent the number of items actually having NC pattern (**h**) but classified as having pattern (**g**).

Clearly

$$\sum_{\mathbf{h}} Z^*_{\mathbf{h}(\mathbf{g})} = Z_{\mathbf{g}}, \quad \text{and} \quad \sum_{\mathbf{g}} Z^*_{\mathbf{h}(\mathbf{g})} = Y_{\mathbf{h}}. \tag{5.32}$$

We will discuss some applications, including grading problems related to results obtained from inspection.

Most of the symbols used in this section are the same as before, but some are new. For easy reference, we have the following list (a few of them have been introduced earlier).

N: the size of the population (lot)
n: the sample size

k: the number of types of nonconformities
$(\mathbf{g})$: an NC pattern, $(g_i = 0, 1)$
$D_{\mathbf{g}}$: the number of items in the lot with NC pattern $(\mathbf{g})$
p_i: probability of correct detection of NC type (i)
p'_i: probability of incorrect detection of NC type (i)
$Y_{\mathbf{g}}$: the number of items in the sample with actual NC pattern $(\mathbf{g})$
$Z_{\mathbf{g}}$: the number of items in the sample judged to have NC pattern $(\mathbf{g})$
$Z^*_{\mathbf{h}(\mathbf{g})}$: the actual number of items among $Z_{\mathbf{g}}$ having NC pattern $(\mathbf{h})$
$\bar{p}_{\mathbf{h}|\mathbf{g}}$: probability that an item with NC pattern $(\mathbf{g})$ will be judged to have pattern $\mathbf{h}$
$p_{\mathbf{h}(\mathbf{g})}$: probability that an item judged to have NC pattern $(\mathbf{g})$ actually has pattern $(\mathbf{h})$
$\mathbf{g}, \boldsymbol{\gamma}$: pattern $(\mathbf{g})$ of inspected types combined with pattern $(\boldsymbol{\gamma})$ of uninspected types

Distributions and moments

We again assume that the results of inspection for each NC type are independent of the presence or absence of other types, and all results for different types are mutually independent, although the *occurrences* of the various types themselves may not be independent. We suppose a sample of size n is randomly chosen from a population (lot) of size N, and, in the lot, there are $D_{\mathbf{g}}$ items having NC pattern $(\mathbf{g})$, for a generic $\mathbf{g}$, so that

$$\sum_{\mathbf{g}} D_{\mathbf{g}} \equiv \sum_{g_1=0}^{1} \cdots \sum_{g_k=0}^{1} D_{g_1,\ldots,g_k} = N. \tag{5.33}$$

The counterpart of the vector $\mathbf{D} \equiv \{D_{\mathbf{g}}\}$ in the sample is the vector $\mathbf{Y} \equiv \{Y_{\mathbf{g}}\}$, in which the 2^k elements are the numbers of items in the sample really having corresponding pattern $(\mathbf{g})$. For instance, the vector $\mathbf{Y}$ in the case $k = 2$ has the form:

$$\mathbf{Y} = (Y_{00}, Y_{01}, Y_{10}, Y_{11}),$$

where the element Y_{ij}, one of $2^2 = 4$, is the number of items in the sample actually having NC pattern (ij).

Because of the independence assumption, the probability that an item having NC pattern $(\mathbf{g})$ will be judged to have NC pattern $(\mathbf{h})$ by inspection is

$$p_{\mathbf{h}|\mathbf{g}} = \prod_{i=1}^{k} p_i^{g_i h_i} p_i'^{(1-g_i)h_i} (1-p_i)^{g_i(1-h_i)} (1-p'_i)^{(1-g_i)(1-h_i)} \tag{5.34}$$

Some numerical values of probabilities of correct decisions, $p_{\mathbf{g}|\mathbf{g}}(=p^{m(\mathbf{g})}(1-p')^{k-m(\mathbf{g})}$ when $p_i=p$, $p'_i=p'$ for all i) are given in Table 5.1(a), ($m(\mathbf{g})=\sum_{i=1}^{k} g_i=$ the number of 1's in ($\mathbf{g}$)).

As before, we first consider distributions conditioned on $\mathbf{Y}=\mathbf{y}$. Similar to (5.33), we have

$$\sum_{\mathbf{g}} Y_{\mathbf{g}}=n \tag{5.35}$$

and also

$$\sum_{\mathbf{g}} Z^{*}_{\mathbf{h}(\mathbf{g})}=Y_{\mathbf{h}}. \tag{5.36}$$

(cf. (5.32)).

Conditional on $\mathbf{Y}=\mathbf{y}$, which means that the number, $y_{\mathbf{h}}$, of pattern ($\mathbf{h}$) items in the sample is given, we have mutually independent binomial variables

$$Z^{*}_{\mathbf{h}(\mathbf{g})} \sim \operatorname{Bin}(y_{\mathbf{h}}, p_{\mathbf{g}|\mathbf{h}}). \tag{5.37}$$

As already noted in (5.32)

$$Z_{\mathbf{g}}=\sum_{\mathbf{h}} Z^{*}_{\mathbf{h}(\mathbf{g})}.$$

Therefore, conditional on $\mathbf{Y}=\mathbf{y}$, the distribution of $Z_{\mathbf{g}}$ is

$$Z_{\mathbf{g}}|\mathbf{Y}=\mathbf{y} \sim *\cdots*_{\mathbf{h}} \operatorname{Bin}(y_{\mathbf{h}}, p_{\mathbf{g}|\mathbf{h}}) \tag{5.38}$$

where $*\cdots*_{\mathbf{h}}$ denotes repeated convolution over all possible 2^k patterns ($\mathbf{h}$) and $p_{\mathbf{g}|\mathbf{f}}$ is given by (5.34).

Conditional on $\mathbf{Y}=\mathbf{y}$, the joint distribution of the $Z_{\mathbf{g}}$'s is thus a convolution of multinomial distributions:

$$\{Z_{\mathbf{g}}\}|\mathbf{y} \sim *\cdots*_{\mathbf{h}} \text{Multinomial}\,(y_{\mathbf{h}}; \{p_{\mathbf{g}|\mathbf{h}}\}) \tag{5.39}$$

Note that, by definition, we have

$$\sum_{\mathbf{g}} p_{\mathbf{g}|\mathbf{h}}=1 \quad \text{and} \quad \sum_{\mathbf{h}} y_{\mathbf{h}}=n. \tag{5.40}$$

The posterior probability that an item actually has pattern ($\mathbf{h}$), given that it is classified as having pattern ($\mathbf{g}$), is

$$p_{\mathbf{h}(\mathbf{g})}=\frac{p_{\mathbf{h}}p_{\mathbf{g}|\mathbf{h}}}{\bar{p}_{\mathbf{g}}}, \tag{5.41}$$

where

$$p_{\mathbf{h}} \equiv \Pr[\text{an item has NC pattern } (\mathbf{h})]=\frac{D_{\mathbf{h}}}{N} \tag{5.42}$$

Table 5.1(a) *Probability* $p_{\mathbf{g}|\mathbf{g}}$ *of correct assignment of* NC *pattern* (**g**) *for* $k = 4$ *types of nonconformity*

$m(\mathbf{g})$	$p' = 0$ $p = 0{\cdot}8$	$p' = 0$ $p = 0{\cdot}95$	$p' = 0$ $p = 1$	$p' = 0{\cdot}05$ $p = 0{\cdot}8$	$p' = 0{\cdot}05$ $p = 0{\cdot}95$	$p' = 0{\cdot}05$ $p = 1$	$p' = 0{\cdot}2$ $p = 0{\cdot}8$	$p' = 0{\cdot}2$ $p = 0{\cdot}95$	$p' = 0{\cdot}2$ $p = 1$
0	1·0000	1·0000	1·0000	0·8145	0·8145	0·8145	0·4096	0·4096	0·4096
1	0·8000	0·9500	1·0000	0·6859	0·8145	0·8574	0·4096	0·4864	0·5120
2	0·6400	0·9025	1·0000	0·5776	0·8145	0·9025	0·4096	0·5776	0·6400
3	0·5120	0·8574	1·0000	0·4864	0·8145	0·9500	0·4096	0·6859	0·8000
4	0·4096	0·8145	1·0000	0·4096	0·8145	1·0000	0·4096	0·8145	1·0000

Probability $p_{\mathbf{g}|\mathbf{g}}$ *of correct assignment of* NC *pattern* (**g**) *for* $k = 6$ *types of nonconformity*

$m(\mathbf{g})$	$p' = 0$ $p = 0{\cdot}8$	$p' = 0$ $p = 0{\cdot}95$	$p' = 0$ $p = 1$	$p' = 0{\cdot}05$ $p = 0{\cdot}8$	$p' = 0{\cdot}05$ $p = 0{\cdot}95$	$p' = 0{\cdot}05$ $p = 1$	$p' = 0{\cdot}2$ $p = 0{\cdot}8$	$p' = 0{\cdot}2$ $p = 0{\cdot}95$	$p' = 0{\cdot}2$ $p = 1$
0	1·0000	1·0000	1·0000	0·7351	0·7351	0·7351	0·2621	0·2621	0·2621
1	0·8000	0·9500	1·0000	0·6190	0·7351	0·7738	0·2621	0·3113	0·3277
2	0·6400	0·9025	1·0000	0·5213	0·7351	0·8145	0·2621	0·3597	0·4096
3	0·5120	0·8574	1·0000	0·4390	0·7351	0·8574	0·2621	0·4390	0·5120
4	0·4096	0·8145	1·0000	0·3697	0·7351	0·9025	0·2621	0·5213	0·6400
5	0·3277	0·7738	1·0000	0·3113	0·7351	0·9500	0·2621	0·6190	0·8000
6	0·2621	0·7351	1·0000	0·2621	0·7351	1·0000	0·2621	0·7351	1·0000

Note the anti-symmetry of the values for ($p' = 0$, $p = 0{\cdot}8$) and ($p' = 0{\cdot}2$, $p = 1$)–values for $m(\mathbf{g}) = m$ in one case equal values for $m(\mathbf{g}) = k - m$ in the other. Also, when $p' = 1 - p$ the values do not depend on $m(\mathbf{g})$.

and

$$\bar{p}_{\mathbf{g}} \equiv \Pr[\text{an item is assigned NC pattern } (\mathbf{g}) \text{ by inspection}]$$

$$= \sum_{\mathbf{h}} p_{\mathbf{h}} p_{\mathbf{g}|\mathbf{h}} = \sum_{\mathbf{h}} p_{\mathbf{g}|\mathbf{h}} \frac{D_{\mathbf{h}}}{N}. \tag{5.43}$$

Table 5.1(b) gives some numerical values of $p_{\mathbf{g}(\mathbf{g})}$.

Notice the difference between the posterior probability $p_{\mathbf{h}(\mathbf{g})}$ and the conditional probability $p_{\mathbf{g}|\mathbf{h}}$, linked by Bayes' formula in (5.41). For example, the distribution of $Z^*_{\mathbf{h}(\mathbf{g})}$, given $Z_{\mathbf{g}} = z_{\mathbf{g}}$ is not necessarily $\text{Bin}(z_{\mathbf{g}}, p_{\mathbf{h}(\mathbf{g})})$ because of the constraints on $Z^*_{\mathbf{h}(\mathbf{g})}$ such as $\sum_{\mathbf{g}} Z^*_{\mathbf{h}(\mathbf{g})} = Y_{\mathbf{h}}$.

The joint distribution of $\mathbf{Y}$ is multivariate hypergeometric with

$$\Pr[\mathbf{Y} = \mathbf{y}] = \left\{ \prod_{\mathbf{g}} \binom{D_{\mathbf{h}}}{y_{\mathbf{g}}} \right\} \bigg/ \binom{N}{n} \tag{5.44}$$

where

$$\prod_{\mathbf{g}} \equiv \prod_{g_1=0}^{1} \cdots \prod_{g_k=0}^{1}.$$

Symbolically,

$$\mathbf{Y} \sim \text{Mult Hypg}(n; \mathbf{D}; N). \tag{5.45}$$

Now $\mathbf{D} \equiv \{D_{\mathbf{g}}\}$ describes the partition of the lot among items having 2^k possible patterns.

Note that there are 2^k different NC patterns and so there are 2^k elements in $\mathbf{Y}$, and in $\mathbf{D}$.

Compounding the conditional distribution (5.39) with the distribution of $\mathbf{Y}$, the overall distribution of $\mathbf{Z}$ is therefore

$$\mathbf{Z} = \{Z_{\mathbf{g}}\} \sim \{* \cdots *_{\mathbf{h}} \text{Multinomial}(Y_{\mathbf{h}}, \{p_{\mathbf{g}|\mathbf{h}}\})\} \bigwedge_{\mathbf{Y}} \text{Mult Hypg}(n; \mathbf{D}; N). \tag{5.46}$$

Or, explicitly

$$\Pr[\mathbf{Z} = \mathbf{z} | n] = \binom{N}{n}^{-1} \sum_{\mathbf{y}} \prod_{\mathbf{h}} D_{\mathbf{h}}^{(y_{\mathbf{h}})} \sum{}^* \prod_{\mathbf{g}} \left\{ \frac{p_{\mathbf{g}|\mathbf{h}}^{z_{\mathbf{g},\mathbf{h}}}}{z_{\mathbf{g},\mathbf{h}}!} \right\}. \tag{5.47}$$

where $\sum^*$ denotes summation with respect to $z_{\mathbf{g},\mathbf{h}}$ subject to

$$\sum_{\mathbf{g}} z_{\mathbf{g},\mathbf{h}} = y_{\mathbf{h}}; \qquad \sum_{\mathbf{h}} z_{\mathbf{g},\mathbf{h}} = z_{\mathbf{g}}.$$

(Note that $z_{\mathbf{g},\mathbf{h}}$ corresponds to a value of $Z^*_{\mathbf{h}(\mathbf{g})}$.)

Table 5.1(b) *Probability* $p_{\mathbf{g}|(\mathbf{g})}$ *that item judged* NC *pattern* (**g**) *actually is* NC *pattern* (**g**) *for* $k = 4$ *types of nonconformity*

$m(\mathbf{g})$	$p' = 0$ $p = 0{\cdot}8$	$p' = 0$ $p = 0{\cdot}95$	$p' = 0$ $p = 1$	$p' = 0{\cdot}05$ $p = 0{\cdot}8$	$p' = 0{\cdot}05$ $p = 0{\cdot}95$	$p' = 0{\cdot}05$ $p = 1$	$p' = 0{\cdot}2$ $p = 0{\cdot}8$	$p' = 0{\cdot}2$ $p = 0{\cdot}95$	$p' = 0{\cdot}2$ $p = 1$
0	0·4823	0·8227	1·0000	0·4646	0·8145	1·0000	0·4096	0·7847	1·0000
1	0·5787	0·8638	1·0000	0·5306	0·8145	0·9524	0·4096	0·6887	0·8333
2	0·6944	0·9070	1·0000	0·6045	0·8145	0·9070	0·4096	0·6045	0·6944
3	0·8333	0·9524	1·0000	0·6887	0·8145	0·8638	0·4096	0·5306	0·5787
4	1·0000	1·0000	1·0000	0·7847	0·8145	0·8227	0·4096	0·4657	0·4823

Probability $p_{\mathbf{g}|(\mathbf{g})}$ *that item judged* NC *pattern* (**g**) *actually is* NC *pattern* (**g**) *for* $k = 6$ *types of nonconformity*

$m(\mathbf{g})$	$p' = 0$ $p = 0{\cdot}8$	$p' = 0$ $p = 0{\cdot}95$	$p' = 0$ $p = 1$	$p' = 0{\cdot}05$ $p = 0{\cdot}8$	$p' = 0{\cdot}05$ $p = 0{\cdot}95$	$p' = 0{\cdot}05$ $p = 1$	$p' = 0{\cdot}2$ $p = 0{\cdot}8$	$p' = 0{\cdot}2$ $p = 0{\cdot}95$	$p' = 0{\cdot}2$ $p = 1$
0	0·3349	0·7462	1·0000	0·3178	0·7351	1·0000	0·2621	0·6951	1·0000
1	0·4019	0·7835	1·0000	0·3621	0·7351	0·9524	0·2621	0·6101	0·8333
2	0·4823	0·8227	1·0000	0·4125	0·7351	0·9070	0·2621	0·5355	0·6944
3	0·5787	0·8638	1·0000	0·4700	0·7351	0·8638	0·2621	0·4700	0·5787
4	0·6944	0·9070	1·0000	0·5355	0·7351	0·8227	0·2621	0·4125	0·4823
5	0·8333	0·9524	1·0000	0·6101	0·7351	0·7835	0·2621	0·3621	0·4019
6	1·0000	1·0000	1·0000	0·6951	0·7351	0·7462	0·2621	0·3178	0·3349

The distribution (5.46) of $\{Z_{\mathbf{g}}\}$, leads to relevant moments

$$E[Z_{\mathbf{g}}) = E_{\mathbf{Y}}[Z_{\mathbf{g}} | \{Y_{\mathbf{h}}\}]$$

$$= \sum_{\mathbf{h}} p_{\mathbf{g}|\mathbf{h}} E[Y_{\mathbf{h}}] = nN^{-1} \sum_{\mathbf{h}} p_{\mathbf{g}|\mathbf{h}} D_{\mathbf{h}} = n\bar{p}_{\mathbf{g}}; \tag{5.48}$$

$$\operatorname{cov}(Z_{\mathbf{g}}, Z_{\mathbf{g}^*}) = -\frac{n(N-n)}{N-1} \bar{p}_{\mathbf{g}} \bar{p}_{\mathbf{g}^*} - \frac{n}{N} \cdot \frac{n-1}{N-1} \sum_{\mathbf{h}} p_{\mathbf{g}|\mathbf{h}} p_{\mathbf{g}^*|\mathbf{h}} D_{\mathbf{h}}; \tag{5.49}$$

$$\operatorname{var}(Z_{\mathbf{g}}) = \operatorname{var}_{\mathbf{Y}}[E[Z_{\mathbf{g}} | \mathbf{Y}]] + E_{\mathbf{Y}}[\operatorname{var}(Z_{\mathbf{g}} | \mathbf{Y})]$$

$$= \frac{n(N-n)}{N(N-1)} \sum_{\mathbf{h}} D_{\mathbf{h}} (p_{\mathbf{g}|\mathbf{h}} - \bar{p}_{\mathbf{g}})^2 + \frac{n}{N} \sum_{\mathbf{h}} D_{\mathbf{h}} p_{\mathbf{g}|\mathbf{h}} (1 - p_{\mathbf{g}|\mathbf{h}}). \tag{5.50}$$

An alternative form of (5.49) is

$$\operatorname{cov}(Z_{\mathbf{g}}, Z_{\mathbf{g}^*}) = -n\{(N-n)\bar{p}_{\mathbf{g}}\bar{p}_{\mathbf{g}^*} + (n-1)\overline{p_{\mathbf{g}}p_{\mathbf{g}^*}}\}(N-1)^{-1}$$

$$= -n\bar{p}_{\mathbf{g}}\bar{p}_{\mathbf{g}^*} - n(n-1)(\overline{p_{\mathbf{g}}p_{\mathbf{g}^*}} - \bar{p}_{\mathbf{g}}\bar{p}_{\mathbf{g}^*})(N-1)^{-1} \tag{5.51}$$

where

$$\overline{p_{\mathbf{g}}p_{\mathbf{g}^*}} = N^{-1} \sum_{\mathbf{h}} D_{\mathbf{h}} p_{\mathbf{g}|\mathbf{h}} p_{\mathbf{g}^*|\mathbf{h}}.$$

Results related to uninspected patterns

It often happens that there are other types $(k+1), \ldots, (k+m)$ of nonconformity that are not inspected. Denote the pattern of these m uninspected types by (γ). We will study distributions and moments relevant to $Z_{\mathbf{g},\gamma}$, which is the number of items classified as having pattern $(\mathbf{g})$ from inspection and actually having uninspected pattern (γ). (We will use Greek letters to denote uninspected patterns.) From our definition,

$$Z_{\mathbf{g}} = \sum_{\gamma} Z_{\mathbf{g},\gamma}, \tag{5.52}$$

and, conditioned on $\mathbf{Y} = \mathbf{y}$,

$$Z_{\mathbf{g},\gamma} | \mathbf{Y} = \mathbf{y} \sim * \cdots *_{\mathbf{h}} \operatorname{Bin}(y_{\mathbf{h},\gamma}, p_{\mathbf{g}|\mathbf{h}}). \tag{5.53}$$

When (γ) and (γ^*) are not identical, $Z_{\mathbf{g},\gamma}$ and $Z_{\mathbf{g}^*,\gamma^*}$ are conditionally independent whether or not $(\mathbf{g})$ and $(\mathbf{g}^*)$ are identical. For identical γ and γ^*, the conditional (on $\mathbf{Y}_\gamma = \mathbf{y}_\gamma$, with $\mathbf{Y}_\gamma = \{Y_{\mathbf{g},\gamma}; \gamma \text{ fixed}\}$) joint distribution of $Z_{\mathbf{g},\gamma}$ and $Z_{\mathbf{g}^*,\gamma}$ with $\mathbf{g} \neq \mathbf{g}^*$ is

$$(Z_{\mathbf{g},\gamma}, Z_{\mathbf{g}^*,\gamma}) | \mathbf{Y}_\gamma = \mathbf{y}_\gamma \sim \\ * \cdots *_{\mathbf{h}} \text{Multinomial}(y_{\mathbf{h},\gamma}; p_{\mathbf{g}|\mathbf{h}}, p_{\mathbf{g}^*|\mathbf{h}}, 1 - p_{\mathbf{g}|\mathbf{h}} - p_{\mathbf{g}^*|\mathbf{h}}). \tag{5.54}$$

The overall joint distribution then can be obtained by compounding with respect to $Y_{\mathbf{h}\gamma}$ (γ fixed)

$$(Z_{\mathbf{g},\gamma}, Z_{\mathbf{g}^*,\gamma}) \sim [* \cdots *_{\mathbf{h}} \text{Multinomial}\,(Y_{\mathbf{h},\gamma}; p_{\mathbf{g}|\mathbf{h}}, p_{\mathbf{g}^*|\mathbf{h}}, 1 - p_{\mathbf{g}|\mathbf{h}} - p_{\mathbf{g}^*|\mathbf{h}})] \\ \bigwedge_{y_{\mathbf{h},\gamma}} \text{Mult Hypg}\,(n; \mathbf{D}_\gamma; N) \tag{5.55}$$

where $\mathbf{D}_\gamma = \{D_{\mathbf{h},\gamma}; \gamma \text{ fixed}\}$. Note that

$$\Pr[\mathbf{Y}_\gamma = \mathbf{y}_\gamma | \gamma \text{ fixed}] = \binom{N}{n}^{-1} \binom{N - D_\gamma}{n - y_\gamma} \prod_{\mathbf{h}} \binom{D_{\mathbf{h},\gamma}}{y_{\mathbf{h},\gamma}}$$

where $y_\gamma = \sum_{\mathbf{h}} y_{\mathbf{h}\gamma}$; $D_r = \sum_{\mathbf{h}} D_{\mathbf{h}\gamma}$.

We will not write down the joint distribution of the $Z_{\mathbf{g},\gamma}$'s since it has the same form as (5.47) for $Z_{\mathbf{g}}$'s, but with each D having a second subscript equal to γ. Similarly, moments and joint moments of $Z_{\mathbf{g},\gamma}$ and $Z_{\mathbf{g}^*,\gamma}$ have the same form as (5.48)–(5.51), with all subscripts $\mathbf{h}$ replaced by $\mathbf{h}, \gamma$ ($\sum_{\mathbf{h}}$ remains unchanged).

By compounding the conditional (independent) joint distribution, with respect to that of $\mathbf{Y}_\gamma$ and $\mathbf{Y}_{\gamma^*}$, which is multivariate hypergeometric:

$$\Pr[\mathbf{Y}_\gamma = \mathbf{y}_\gamma; \mathbf{Y}_{\gamma^*} \equiv \mathbf{y}_{\gamma^*}] = \binom{N}{n}^{-1} \binom{N - D_\gamma - D_{\gamma^*}}{n - \mathbf{y}_\gamma - \mathbf{y}_{\gamma^*}} \prod_{\mathbf{h}} \left\{ \binom{D_{\mathbf{h},\gamma}}{\mathbf{y}_{\mathbf{h},\gamma}} \binom{D_{\mathbf{h},\gamma^*}}{\mathbf{y}_{\mathbf{h},\gamma^*}} \right\} \tag{5.56}$$

we obtain the overall joint distribution of $Z_{\mathbf{g},\gamma}$ and $Z_{\mathbf{g}^*,\gamma^*}$ ($\gamma \neq \gamma^*$). Using the mutual independence between $Z_{\mathbf{g},\gamma}$ and $Z_{\mathbf{g}^*,\gamma^*}$, conditioned on $\mathbf{Y}_\gamma = \mathbf{y}_\gamma$, we obtain

$$\text{cov}(Z_{\mathbf{g},\gamma}, Z_{\mathbf{g}^*,\gamma^*}) = -\frac{n(N-n)}{N^2(N-1)} \left(\sum_{\mathbf{h}} D_{\mathbf{h},\gamma} p_{\mathbf{g}|\mathbf{h}} \right) \left(\sum_{\mathbf{h}} D_{\mathbf{h},\gamma^*} p_{\mathbf{g}^*|\mathbf{h}} \right) \tag{5.57}$$

(cf. (5.49)).

This must be negative, reflecting the fact that $Z_{\mathbf{g}}$ and $Z_{\mathbf{g}^*}$ are negatively correlated (since the more ($\mathbf{g}$) decisions, the fewer remaining decisions that might be ($\mathbf{g}^*$)).

Before concluding this chapter we note once more the paper by Bonett (1988). This deals with estimation of the numbers of nonconformities of different types, using a contingency table approach. He uses the number of 'defects' of various kinds identified as such by several different inspectors. This is broadly related to an approach to group testing using repeated observations developed by Graff and Roellofs (see Chapter 7).

PART TWO

Identification of nonconforming items

CHAPTER 6

(Dorfman) group testing

Years ago, old Mr Lyle carried a milk bottle, filled to the brim with yellow fluid, to the Ballantrae Polyclinic and Health Centre. He went into the clinic and handed the bottle carefully to a nurse. 'Here it is, Nurse Cameron. The urine specimen. Name of Fergus Lyle...'

'Aye. Coom back in three days for the analysis.'

In three days, Mr Lyle returned. He was taken to see Dr Howard, a young doctor. The doctor glanced down a report and said, 'Well, Mr Lyle, the laboratory report they gie ye is all fine. Complete bill o' health. No diseases.'

Mr Lyle went home with great speed and scarcely contained delight. Before he was through the door of his house he cried, 'Kathy, lass, Kathy! It's great news I am bringin'!'

His wife hurried in from the kitchen, 'Aye?'

'Hoot, Kathy, it's fine I am in all respects! And so are ye, and so is y'r old dad an' y'r brother Nairn, and bonnie cousin Molly, and also me brother Clyde, an' his woman, an' my old friend Douglas, an' th' charwoman...' (Rosten, L. (1985) *Giant Book of Laughter*, p. 463, Crown, New York.)

6.1 Introduction

In 1943 the *Annals of Mathematical Statistics* contained a short note by Robert Dorfman (then a senior statistician in the Research Division of the Office of Price Administration, and currently a Professor of Economics at Harvard University) explaining an efficient method for eliminating all defective (NC) items from a certain large population. The wartime draft used Wasserman screening. The note, based on an idea of Dorfman and Murray Rosenblatt, suggested simplifying the testing of draftee blood samples by using probability rather than doing millions of tests. This short note revolutionized the field of testing of defective items and has resulted in an enormous number of publications, mainly – though not exclusively – by M. Sobel and his associates. This field is still very active today, after almost half a century. A review paper by A.B. Mundel in the *Journal of Quality Technology* (1984) provides a summary of more recent developments. Because of its importance and influence, Dorfman's paper is reproduced as Appendix B. The paper describes, in some detail, an application of his method to weeding out all syphilitic men among those called up for induction into the US Armed Forces in the years 1941–1943. Instead of testing each blood sample individually, Dorfman proposed to pool n_0 samples in a single analysis. Presence of syphilitic antigen in the pool leads to a decision to make n_0 individual tests; absence of such an antigen leads to immediate clearance of all n_0 men without further testing. In his note, Dorfman presents numerical results of application of his procedure to the first million selectees and volunteers. He was mainly concerned with savings expected to result from application of his procedure. This depends on the group size (n_0) and the prevalence rate of the disease. If the latter is known, it is possible to choose an optimal group size, maximizing the expected savings in testing.

On the assumption that testing is error-free, for a prevalence rate of 5% the optimum size of the group is 5 and the percent expected savings is 57%; for a rate of 1% the optimum size is 11 and the expected saving skyrockets to 80%. When the prevalence rate is 30%, however, the optimum size decreases to 3 and the expected savings is barely 1%. Dorfman (1943) emphasized that his group testing technique can economically be applied only under the following conditions:

1. The prevalence rate (i.e. the proportion, ω, of NC items) is sufficiently small.
2. It is easier, or at least as economical, to obtain an observation on a group as on an individual of the group tested separately. Dorfman remarks that this procedure can also be applied to situations where the interest centres on the degree to which an 'imperfection' is present rather than merely in its presence or absence, by regarding amounts of imperfection in excess of a specific level as 'nonconformity', otherwise as 'conformity'.

Of course, the original analysis presented by Dorfman assumes that inspection is error-free ($p = 1$; $p' = 0$).

It should be kept in mind that group testing methods can be used only when it is possible to determine whether a group of, say, n_0 items contains any nonconforming (NC) items by a single test on the group as a whole. Individual testing therefore is a special case of group testing with $n_0 = 1$. For example, 'items' might be containers of liquid and 'nonconformity' the presence of a specified contaminant. Mixing liquid from each container into a group, one can test the mixture for presence of contaminant. Other examples can be found in a quite extensive literature – including Chen and Swallow (1990), Hwang (1984), Mundel (1984), Sobel (1968), Swallow (1985, 1987), Thomas *et al.* (1973, 1975), Watson (1961) and Wolf (1985).

The original Dorfman screening procedure is as follows. A group test is applied first. (i) If the group test gives a positive result, indicating the existence of at least one NC item in the group, each item is tested individually. (ii) If the group test has a negative result, all items in the group are declared C, and no more tests are needed. (This is illustrated in Fig. 6.1.) Obviously, there may be substantial savings in the expected number of tests if the proportion of NC items in the population is small.

There are modified Dorfman procedures, which may mean bigger savings in the number of tests. For example, instead of testing individual items after a positive result from a group test, subgroups are tested, and this procedure is repeated and continued until all

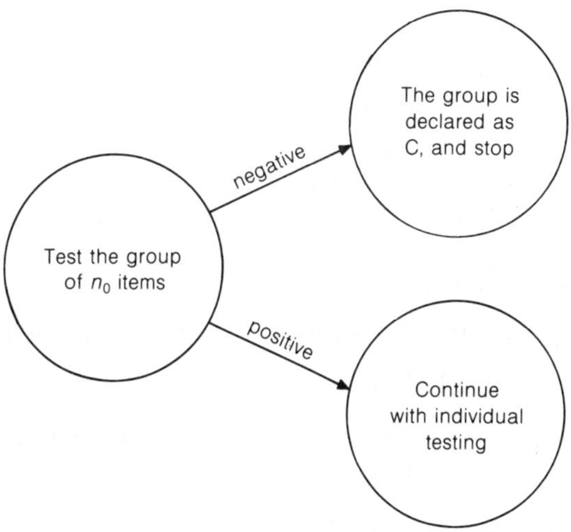

Figure 6.1 *Flow chart for Dorfman procedure*

items are classified. This is **hierarchal screening** which we will discuss in section 6.3.

Section 6.4 will consider the case of testing for two types of nonconformity, A and B, where individual tests for one type of nonconformity A, say, are made first. When a certain number of NC decisions has been made, a Dorfman group testing procedure for type B is applied to those tested items which were not declared to have A type nonconformity. We therefore have a regular Dorfman procedure for NC type B with a random group size, which depends on the number of items with NC type A. This scheme was proposed by Kemp and Kemp (1988). The analysis in this section is concerned with the magnitude of the effects of errors in inspection testing, with particular reference to the Kemp and Kemp scheme.

Three quantities: PC(NC), PC(C) and E, which are the probabilities of correctly identifying an NC item, a C item, and the expected number of tests, respectively, play important roles in measuring the effectiveness of our procedures.

A different application of group testing is described in Gastwirth and Hammick (1989) and Hammick and Gastwirth (1990). Here the aim is to estimate ω without identifying any individual item as NC or C.

6.2 Original Dorfman procedure

Notation

Suppose a sample of size n_0 is taken from a lot of size N. (Here we use n_0, instead of n, because it is convenient when extending the notation to hierarchal screening, where division into subgroups may be of interest.) Corresponding to p and p' for individual testing, we use p_0 and p'_0 to denote the probabilities of a correct positive result and an incorrect positive result, respectively, for testing a group of size n_0. Symbolically,

$$p_0 \equiv \Pr[W=1 \mid Y>0]; \qquad p'_0 \equiv \Pr[W=1 \mid Y=0], \tag{6.1}$$

where the variable W is an indicator, defined by

$$W = \begin{cases} 1 & \text{if the group test gives a positive result;} \\ 0 & \text{otherwise.} \end{cases} \tag{6.2}$$

(As mentioned before, 'a positive result' in group testing means an indication of the existence of at least one NC item in the group.)

We will still use Z to represent the number of items declared NC on individual testing, and D and Y for the numbers of truly NC items in the lot and the sample, respectively.

Analysis

The probability of obtaining a positive result (whether it is a correct or incorrect NC decision) on the group test is

$$\{1-\mathbf{p}_0(n_0)\}p_0 + P_0(n_0)p'_0 = p_0 - (p_0 - p'_0)P_0(n_0), \tag{6.3}$$

where

$$P_0(n_0) = \frac{(N-D)^{(n_0)}}{N^{(n_0)}}$$

is the probability that the sample of size n_0 contains no NC items (i.e. $Y=0$).

It is clear that if $W=0$, there is no need for further inspection, and if $W=1$, the subsequent individual testing is the same as for the single sampling scheme discussed in Chapter 2.

We then have the distribution of Z, symbolically

$$Z \sim [\text{Bin}(WY, p_0) * \text{Bin}(W(n_0 - Y), p'_0)] \bigwedge_Y \text{Hypg}(n_0; D, N), \tag{6.4a}$$

and explicitly

$$\Pr[Z=z] = E_{W,Y}\Bigg[\sum_x \binom{WY}{x} p_0^x (1-p_0)^{WY-x} \times \binom{W(n_0-Y)}{z-x} p_0'^{z-x}(1-p'_0)^{W(n_0-Y)-z+x}\Bigg], \tag{6.4b}$$

where the summation $\sum_x$ is taken over the range $\max(0, z - W(n_0 - Y)) \leqslant x \leqslant \min(WY, z)$.

We start with conditional factorial moments

$$E[Z^{(r)} \mid W, Y] = \sum_{i=0}^{r} \binom{r}{i} (WY)^{(i)} \{W(n_0 - Y)\}^{(r-i)} p^i p'^{r-i}. \tag{6.5}$$

Note that

$$E[Y^{(i)}(n_0 - Y)^{(r-i)} \mid Y > 0] \Pr[Y>0] = \begin{cases} n_0^{(r)} D^{(i)} (N-D)^{(r-i)} / N^{(r)}, & \text{for } i > 0 \\ n_0^{(r)}[(N-D)^{(r)}\{N^{(r)}\}^{-1} - P_0(n_0)], & \text{for } i = 0 \end{cases} \tag{6.6}$$

and

$$E[Y^{(i)}(n_0 - Y)^{(r-i)} | Y = 0] \Pr[Y = 0] = \begin{cases} 0, & \text{for } i > 0 \\ n_0^{(r)} P_0(n_0) & \text{for } i = 0 \end{cases}. \tag{6.7}$$

From (6.1), (6.2), (6.6) and (6.7), we find the factorial moments

$$E[Z^{(r)}] = n_0^{(r)} \left\{ p_0 (N^{(r)})^{-1} \sum_{i=0}^{r} \binom{r}{i} D^{(i)} (N - D)^{(r-i)} p^i p'^{(r-i)} - (p_0 - p_0') p'^r P_0(n_0) \right\}. \tag{6.8}$$

In particular, for $r = 1$, the mean is

$$E[Z] = n_0 \{ p_0 \bar{p} - (p_0 - p_0') p' P_0(n_0) \} = n_0 (p_0 \bar{p} - P_{(0)} p'), \tag{6.9}$$

where

$$\bar{p} = N^{-1} \{ Dp + (N - D) p' \}$$

and

$$P_{(0)} \equiv (p_0 - p_0') P_0(n_0),$$

and the variance is

$$\operatorname{var}(Z) = n_0^{(2)} \left[p_0 \left\{ \bar{p}^2 - \frac{D(p^2 - \bar{p}^2) + (N - D)(p'^2 - \bar{p}^2)}{N(N-1)} \right\} - P_{(0)} p'^2 \right] + n_0 (p_0 \bar{p} - P_{(0)} p') - n_0^2 (p_0 \bar{p} - P_{(0)} p')^2. \tag{6.10}$$

As a special case, if $p_0' = p' = 0$, i.e. there is no possibility of 'false positive', so that all NC decisions for either group or individuals are correct (but there might be incorrect C decisions), we have

$$E[Z] = \frac{n_0 p_0 p D}{N},$$

$$\operatorname{var}(Z) = \frac{n_0 p_0 p D}{N} \left(1 - \frac{p_0 p D}{N} \right) + \frac{n_0 (n_0 - 1) p D}{N^2 (N - 1)} \times \{ N(D - 1) - (N - 1) D p_0 \}.$$

It can usually be expected that $p > p'$, otherwise the tests would be totally unreliable. We can also expect that $p_0 > p_0'$, for the same reason. Because group testing may be more difficult, it is possible that $p_0 < p$. If p_0 depends on the value of Y, (possibly increasing with Y), the distribution of Z will be more complicated.

The three quantities suitable for measuring the effectiveness of the group testing are:

1. probability of correct classification for an NC item, which is

$$\mathrm{PC(NC)} = p_0 p; \tag{6.11}$$

2. probability of correct classification for a C item, which is

$$\begin{aligned}\mathrm{PC(C)} &= P_0^*(n_0)(1 - p_0' p') + (1 - P_0^*(n_0))(1 - p_0 p') \\ &= 1 - (p_0 - P_{(0)}^*)p',\end{aligned} \tag{6.12}$$

   where

$$P_{(0)}^* \equiv (p_0 - p_0')P_0^*(n_0);$$

   and $\qquad P_0^*(n_0) \equiv (N - D - 1)^{(n_0 - 1)}/(N - 1)^{(n_0 - 1)},$

   which is the probability that all items in the sample of size n_0 are C, given that a *specified* one of them is C (cf. the definitions of $P_{(0)}$ and P_0 (n_0 in (6.3)); and
3. the expected number of tests is

$$\begin{aligned}E &= 1 + n_0[P_0(n_0)p_0' + \{1 - P_0(n_0)\}p_0] \\ &= 1 + n_0(p_0 - P_{(0)}).\end{aligned} \tag{6.13a}$$

Clearly, large values of PC(C) and PC(NC), and small values of E, would indicate good procedures.

Since $\mathrm{PC(NC)} = p_0 p \leqslant p$, it is impossible to increase PC(NC). But the screening does increase PC(C), the probability of correctly indicating a C item, and usually reduces E, the expected number of tests.

Without group testing, the expected number of tests is n_0. The expected proportional average saving on the number of tests from screening is

$$1 - E/n_0 = 1 - \{1 + n_0(p_0 - P_{(0)})\}/n_0 = 1 - n_0^{-1} - p_0 + P_{(0)}, \tag{6.13b}$$

which is called EPR (expected proportional reduction).

In the case $N \to \infty$, we consider $D \to \infty$, and $\omega = D/N$ remains unchanged.

Table 6.1 is Table 1 of Kotz and Johnson (1982), and provides

Table 6.1 *Simple screening*

	ω	n_1	PC(C) = *Probability of correct classification of nonconforming*	EPR = *Expected percent reduction in tests* (%)
$p_0 = 0{\cdot}98$, $p'_0 = 0{\cdot}05$, $p' = 0{\cdot}05$	0·05	6	0·9870	53·7
		8	0·9835	51·2
		10	0·9803	47·7
		12	0·9774	43·9
	0·1	6	0·9785	34·8
		8	0·9732	29·5
		10	0·9690	24·4
		12	0·9656	20·9
	0·2	6	0·9662	9·7
		8	0·9608	5·1
		10	0·9572	2·0
		12	0·9550	0·0
$p_0 = 0{\cdot}95$, $p'_0 = 0{\cdot}05$, $p' = 0{\cdot}05$	0·05	6	0·9873	54·5
		8	0·9839	52·2
		10	0·9804	48·9
		12	0·9781	45·3
	0·1	6	0·9791	36·2
		8	0·9740	31·2
		10	0·9699	26·4
		12	0·9666	22·1
	0·2	6	0·9672	11·9
		8	0·9619	7·6
		10	0·9589	4·7
		12	0·9564	2·9
$p_0 = 0{\cdot}90$, $p'_0 = 0{\cdot}05$, $p' = 0{\cdot}05$	0·05	6	0·9878	55·8
		8	0·9847	53·9
		10	0·9818	50·9
		12	0·9792	47·6
	0·1	6	0·9801	38·5
		8	0·9753	34·1
		10	0·9715	29·6
		12	0·9683	25·7
	0·2	6	0·9689	15·5
		8	0·9639	11·8
		10	0·9607	9·1
		12	0·9587	7·5

Table 6.1 (*Contd.*)

	ω	n_1	PC(C) = *Probability of currect classification of nonconforming*	EPR = *Expected percent reduction in tests (%)*
$p_0 = 0{\cdot}95$, $p'_0 = 0{\cdot}10$, $p' = 0{\cdot}10$	0·05	6	0·9718	60·8
		8	0·9644	58·9
		10	0·9586	55·9
		12	0·9533	52·6
	0·1	6	0·9552	43·5
		8	0·9457	39·1
		10	0·9379	34·6
		12	0·9317	30·7
	0·2	6	0·9329	20·5
		8	0·9228	16·8
		10	0·9164	14·1
		12	0·9123	12·5

Note that in the last set, the expected percent reduction in expected number of tests is always 5% greater than for the corresponding case in the penultimate set. This is because the value of $p_0 - p'_0$ is the same (0·85) in the two sets while the value of p_0 is 0·05 greater in the last set.

values of PC(C) and EPR $= (1 - E/n_0)$ for sets of parameters values:

$$(p_0, p'_0, p') = (0{\cdot}98,\ 0{\cdot}05,\ 0{\cdot}05),\ (0{\cdot}95,\ 0{\cdot}05,\ 0{\cdot}05),\ (0{\cdot}90,\ 0{\cdot}05,\ 0{\cdot}05),\ (0{\cdot}95,\ 0{\cdot}10,\ 0{\cdot}10);$$

$$\omega = 0{\cdot}05,\ 0{\cdot}1,\ 0{\cdot}2; \qquad n_0 = 6,\ 8,\ 10,\ 12.$$

It is to be noted that, as might be expected, screening increases PC(C) and substantially reduces E. (Note that the values shown do not depend on p, but PC(NC) $= p_0 p$.)

6.3 Hierarchal screening

Introduction

For screening a group of size n_0, we consider the following procedure. After obtaining a positive result in the first-stage group testing, instead of testing individual items as in the standard Dorfman procedure, we test subgroups, and after each subgroup testing in the second stage, if we obtain a positive result, we repeat the same procedure again. Of course, we will stop testing those subgroups from which we get a negative result. This idea leads to a multi-stage hierarchal Dorfman procedure. In some circumstances, greater savings in expected numbers of tests can be attained by applying this kind of procedure.

We denote the original group of size n_0 by sub^{0}group, and it is divided into h_1 subgroups of size n_1, each called a sub^{1}group. We continue this dividing process up to h_k sub^{k}groups each of size n_k in each sub^{k-1}group. We use 'sub^{k+1}group' to denote individual items. Generally each sub^{j-1}group contains h_j sub^{j}groups, of size n_j each, so that $n_j = n_{j-1}/h_j$; $j = 1, \ldots, k+1$.

Formally, the corresponding $(k+1)$-stage hierarchal Dorfman procedure is as follows. First, the group of n_0 items is tested as a whole, as in the standard Dorfman procedure. If no NC items are indicated, all items are declared conforming (C). However, if the result is positive, i.e. the group is declared NC, each subgroup (of $n_1 = n_0/h_1$ items) is tested as a whole. Generally, when a sub^{j}group (of $n_j = n_0/(\prod_{i=1}^{j} h_i)$ items) is tested, all items in the sub^{j}group are declared C if the group testing has a negative result; otherwise each of the h_{j+1} sub^{j+1}groups (of $n_{j+1} = n_j/h_{j+1}$ items) in the sub^{j}group is tested as a whole.

Continuing in this way, if there is an NC result from testing a sub^{k}group, which is the last real subgroup, then each of the

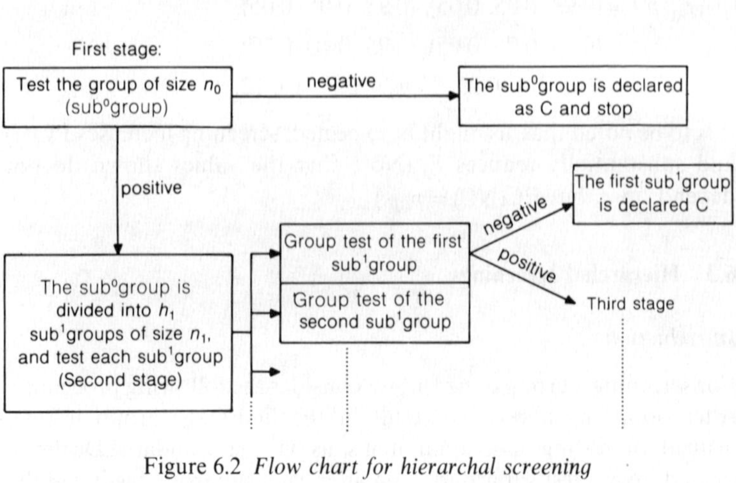

Figure 6.2 *Flow chart for hierarchal screening*

$n_{k+1}(=n_0/(\prod_{j=1}^{k+1}h_j))$ individual items ('sub^{k+1}groups') contained in the sub^{k}group is tested individually. (Note: $h_{k+1}=n_k$.) In other words, the test of the sub^{k}group is the same as standard Dorfman testing. (This is illustrated in Fig. 6.2.)

Hierarchal screening procedures are discussed by Kotz and Johnson (1982b). They analyse the effects of imperfect inspection on three-stage procedures (i.e. $k=2$) and provide some tables.

There are examples of more elaborate hierarchal procedures in Hwang (1984). Interesting relationships between screeening procedures and multi-access communication theory are discussed in Wolf (1985).

Analysis

We now present formulas for properties of a $(k+1)$-stage hierarchal procedure. These formulas should be compared with their analogues, obtained in previous sections. For sub^{s}group testing, we define symbols p_s and p'_s by

$$\begin{aligned} p_s &= \Pr[\text{declare sub}^s\text{group NC}|\text{at least one NC item}] \\ p'_s &= \Pr[\text{declare sub}^s\text{group NC}|\text{no NC items}], \end{aligned} \tag{6.14}$$

for $s=0,\ldots,k+1$. One may notice that (6.14) replaces p and p' by p_{k+1} and p'_{k+1}, respectively, for simplicity and compactness of formulas. We will still use the symbols, for $i=0,1,\ldots,k$,

$$P_0(n_i)\equiv\frac{(N-D)^{(n_i)}}{N^{(n_i)}},$$

the probability that a sample of size n_i contains no NC items; and

$$P_{(i)}\equiv(p_i-p'_i)P_0(n_i).$$

We also define, for $i=0,1,\ldots,k$,

$$P_0^*(n_i)\equiv\frac{(N-D-1)^{(n_i-1)}}{(N-1)^{(n_i-1)}},$$

which is the probability that all items in the sample of size n_i are C, given that a *specified* one of them is C, and

$$P_{(i)}^*\equiv(p_i-p'_i)P_0^*(n_i).$$

For an infinite lot size with proportion ω of NC items

$$P_0(n_i) \equiv (1-\omega)^{n_i}; \qquad P_0^*(n_i) \equiv (1-\omega)^{n_i - 1}.$$

We now calculate the three quantities measuring the effectiveness of the screening. The probabilities of correct classifications (for NC and C items, respectively) are

$$\mathrm{PC(NC)} = \prod_{j=0}^{k+1} p_j; \tag{6.15}$$

$$\mathrm{PC(C)} = 1 - \sum_{s=0}^{k+1} \{P_0^*(n_s) - P_0^*(n_{s-1})\}\left(\prod_{j=0}^{s-1} p_j\right)\left(\prod_{j=s}^{k+1} p_j'\right), \tag{6.16}$$

with $P_0^*(n_{-1}) = 0$ and $P_0^*(n_{k+1}) = 1$.

The expected number of tests is

$$E = \sum_{s=0}^{k+1}\left(\prod_{j=1}^{s} h_j\right)\pi(s), \tag{6.17}$$

where

$$\pi(s) = P_0(n_0)\prod_{i=0}^{s-1} p_i' + \sum_{j=1}^{s-1}\{P_0(n_j) - P_0(n_{j-1})\}$$

$$\times\left(\prod_{i=0}^{j-1} p_i\right)\left(\prod_{i=j}^{s-1} p_i'\right) + \{1 - P_0(n_{s-1})\}\prod_{i=0}^{s-1} p_i, \tag{6.18}$$

with $\prod_{j=1}^{0} h_j = \pi(0) = 1$ and $p'_{-1} = p_{-1} = 0$, or in words;

$\pi(s) = \Pr$[a randomly chosen sub^{s+1}group will be tested and judged NC].

Note that $\prod_{j=1}^{s} h_j$ is the number of sub^{s+1}groups.

The details of the derivation of the above formulas can be found in section 9.3 (for almost identical situations).

The ratio of the expected number of tests (E) to the number of items ($n_0 = \prod_{j=1}^{k+1} h_j$) is

$$E/n_0 = \sum_{s=0}^{k+1}\left(\prod_{j=s+1}^{k+1} h_j\right)^{-1}\pi(s). \tag{6.19}$$

The simplest case, other than the standard Dorfman procedure which is two-stage ($k=1$), is a three-stage hierarchal procedure ($k=2$). The formulas for the three quantities, PC(NC), PC(C) and E, when $k=2$,

Table 6.2 *Values of* PC(C) *and* EPR *with* $n_0 = 12$; $p_0 = p = 0{\cdot}9$

Test procedure		$p'_0 = p' = 0{\cdot}05$				$p'_0 = p' = 0{\cdot}1$			
		$\omega = 0{\cdot}05$		$\omega = 0{\cdot}1$		$\omega = 0{\cdot}05$		$\omega = 0{\cdot}1$	
		PC(C)	EPR(%)	PC(C)	EPR(%)	PC(C)	EPR(%)	PC(C)	EPR(%)
Simple Dorfman		0·9792	47·6	0·9801	25·7	0·9555	44·7	0·9351	24·3
Three-stage Dorfman									
n_1	n_2								
4	2	0·9980	64·9	0·9960	45·8	0·9953	62·8	0·9909	43·6
6	2	0·9978	65·3	0·9957	46·1	0·9947	63·3	0·9899	43·8
6	3	0·9962	65·9	0·9926	47·0	0·9916	64·3	0·9841	45·3

are

$$\mathrm{PC(NC)} = p_0 p_1 p_2; \tag{6.20}$$

$$\begin{aligned}\mathrm{PC(C)} &= P_0^*(n_0)(1 - p_0' p_1' p_2') + (1 - P_0^*(n_0))(1 - p_0) \\ &\quad + (P_0^*(n_1) - P_0^*(n_0))p_0(1 - p_1' p_2') + (1 - P_0^*(n_1))p_0(1 - p_1 p_2') \\ &= 1 - p_0 p_1 p_2' + P_0^*(n_0)(p_0 - p_0')p_1' p_2' + P_0^*(n_1)(p_1 - p_1')p_0 p_2' \\ &= 1 - p_0 p_1 p_2' + (P_{(0)}^* p_1' + P_{(1)}^* p_0)p_2';\end{aligned} \tag{6.21}$$

and

$$\begin{aligned}E &= 1 + h(p_0 - P_{(0)}) \\ &\quad + n_0[P_0(n_0)p_0' p_1' + (1 - P_0(n_1))p_0 p_1 + (p_0(n_1) - P_0(n_0))p_0 p_1'] \\ &= 1 + h(p_0 - P_{(0)}) + n_0(p_0 p_1 - P_{(0)} p_1' - P_{(1)} p_0).\end{aligned} \tag{6.22}$$

The proportional reduction in the expected number of tests (EPR) is

$$1 - E/n_0 = 1 - n_0^{-1} - p_0 p_1 - n_1^{-1}(p_0 - P_{(0)}) + P_{(0)} p_1' + P_{(1)} p_0. \tag{6.23}$$

Table 8 of Johnson *et al.* (1988) gives values of PC(NC), PC(C), E, and EPR ($= 1 - E/n_0$) for some three-stage procedures with several sets of parameters. Comparing with standard Dorfman procedures, Tables 6 and 8 of Johnson *et al.* (1988) show the savings in expected testing by applying hierarchal schemes.

We present, here, comparisons between some representative three-stage procedures and simple Dorfman procedures (Table 6.2) having the same total group size ($n_0 = 12$). In all cases, the probability of correct classification for an NC group, subgroup or item ($p_0 = p$) is 0·9. The calculations are based on an 'infinite' lot size with a proportion ω of NC items.

For an infinite lot size, E increases with ω, but PC(C) does not depend on ω very much, while it increases with n_0. Note that PC(C) is increased greatly by application of screening procedures.

As one might expect, further savings are obtained when the group sizes encountered are such that savings would be obtained with simple Dorfman procedures applied to those group sizes.

For a fair assessment of the improvement due to hierarchal screening we should use a more nearly optimal value of n_0 with the simple Dorfman procedure. For the case $p = 0·9$, $p' = 0·05$ with $n_0 = 6$ we get EPR = 38·5% while with three-stage hierarchal screening with $n_0 = 12$, $n_1 = 6$ we get EPR = 46·1% or 47·0% depending on the size of sub²groups (2 or 3, respectively). (PC(C) is 0·9801 for the simple procedure, 0·9957 or 0·9826 for the hierarchal screening procedures.)

6.4 Dorfman procedures with random group size and two NC types

Introduction

The scheme presented here was proposed by Kemp and Kemp (1988). Assume the lot size is effectively infinite, with two types of nonconformity, A and B, and denote proportions of items with these two NC types in the lot by ω_A and ω_B, respectively. We also assume that A and B occur independently in the lot and that results of tests are mutually independent.

The procedure is as follows. First, we carry out sequential individual inspection for type A until k *successive* items with A are observed (cf. CSP-1, Section 4.5). Let the total number of items so far tested be $k + N$. The k items observed with A are then discarded. The remaining N (a random variable) items will be tested for B using the Dorfman group screening procedure, i.e. they are first tested as a group, and then tested individually if and only if the group testing has a positive result. (This is illustrated Fig. 6.3)

The motivation for this procedure is that A is relatively easy to inspect, while B is more important, but rarer and more difficult to detect. The proposed procedure is aimed at reducing interruptions in production runs required to apply Dorfman procedures.

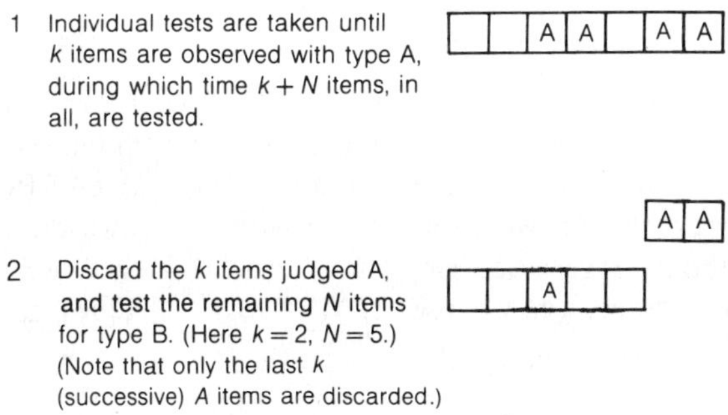

Figure 6.3 *Kemp and Kemp's procedure*

Notation

For G ≡ A, B, the two types of nonconformity, let

p_G = Pr [a correct G-type NC decision is made on individual testing];
p'_G = Pr [an incorrect G-type NC decision is made on individual testing].

Since group testing is used only for type B, we define

p_{OB} = Pr [a correct B-type NC decision is made on group testing];
p'_{OB} = Pr [an incorrect B-type NC decision is made on group testing].

If inspection is perfect, we would have

$$p_A = p_B = p_{OB} = 1; \qquad p'_A = p'_B = p'_{OB} = 0.$$

For the Dorfman procedure used here, symbols PC(NC), PC(C) and E have the same meanings as in previous sections.

B-type nonconformity

For fixed N, and B-type nonconformity, it follows from (6.11), (6.12) and (6.13), (noting that $P_0^*(N) = (1-\omega)^{N-1}$ and $P_0(N) = (1-\omega)^N$ for infinite lot size) that

$$\text{PC(NC)} = p_{OB}p_B \qquad \text{(for any } N\text{)}; \tag{6.24a}$$
$$\text{PC(C)} = 1 - p_{OB}p'_B + (p_{OB} - p'_{OB})p'_B(1-\omega_B)^{N-1}; \tag{6.24b}$$
$$E = 1 - N\{p_{OB} - (p_{OB} - p'_{OB})(1-\omega_B)^N\}. \tag{6.24c}$$

Since N is now random, we take expectations of the above quantities with respect to N, obtaining,

$$\text{PC(NC)} = p_{OB}p_B \tag{6.25a}$$
$$E = 1 + p_{OB}E[N] - (p_{OB} - p'_{OB})E[N(1-\omega_B)^N]. \tag{6.25b}$$

In the last term of the expression for PC(C) in (6.24b), the factor $(1-\omega_B)^{N-1}$ decreases with N, and therefore, after taking expectation with respect to N, we would obtain an average value of PC(C) per Dorfman procedure, which decreases with N. However, what we want is the average value *per C item* instead. Hence, we take the weighted average, i.e. we use $E[N(1-\omega_B)^{N-1}]/E[N]$ to replace $(1-\omega_B)^{N-1}$. Accordingly,

$$\text{PC(C)} = 1 - p_{OB}p'_B + (p_{OB} - p'_{OB})p'_B\frac{E[N(1-\omega_B)^{N-1}]}{E[N]}. \tag{6.25c}$$

From the generating function for N given by Kemp and Kemp (1988),

$$G_N(s) = \sum_{n=\infty}^{\infty} \Pr[N=n]s^n = \tilde{\omega}_{\mathrm{A}}^k(1-\tilde{\omega}_{\mathrm{A}})\{1-s+(1-\tilde{\omega}_{\mathrm{A}}s)\tilde{\omega}_{\mathrm{A}}^k s^{k+1}\}^{-1}, \tag{6.26}$$

where $\tilde{\omega}_{\mathrm{A}} = \omega_{\mathrm{A}}p_{\mathrm{A}} + (1-\omega_{\mathrm{A}})p'_{\mathrm{A}}$ is the probability that a randomly chosen item is indicated by inspection, to have A-type nonconformity we obtain

$$E[N] = G'_N(s)|_{s=1}; \tag{6.27a}$$

$$E[N(1-\omega_{\mathrm{B}})^{N-1}] = G'_N(s)|_{s=1-\omega_{\mathrm{B}}}; \tag{6.27b}$$

$$E[N(1-\omega_{\mathrm{B}})^{N}] = (1-\omega_{\mathrm{B}})G'_N(s)|_{s=1-\omega_{\mathrm{B}}}. \tag{6.27c}$$

Since

$$G'_N(s) = \frac{(1-\tilde{\omega}_A)\tilde{\omega}_A^k}{\{1-s+(1-\tilde{\omega}_A)\tilde{\omega}_A^k s^{k+1}\}^2}\{1-\tilde{\omega}_A^k s^k(k+1-k\tilde{\omega}_A s)\}, \tag{6.28}$$

it follows that

$$E[N] = \frac{1-\tilde{\omega}_{\mathrm{A}}^k(k+1-k\tilde{\omega}_{\mathrm{A}})}{(1-\tilde{\omega}_{\mathrm{A}})\tilde{\omega}_{\mathrm{A}}^k}; \tag{6.29a}$$

$$E[N(1-\omega_{\mathrm{B}})^{N-1}] = \frac{(1-\tilde{\omega}_{\mathrm{A}})\tilde{\omega}_{\mathrm{A}}^k}{\{\omega_{\mathrm{B}}+(1-\tilde{\omega}_{\mathrm{A}})\tilde{\omega}_{\mathrm{A}}^k(1-\omega_{\mathrm{B}})^{k+1}\}^2} \times [1-\tilde{\omega}_{\mathrm{A}}^k(1-\omega_{\mathrm{B}})^k\{k+1-k\tilde{\omega}_{\mathrm{A}}(1-\omega_{\mathrm{B}})\}]; \tag{6.29b}$$

$$E[N(1-\omega_{\mathrm{B}})^{N}] = (1-\omega_{\mathrm{B}})E[N(1-\omega_{\mathrm{B}})^{N-1}]. \tag{6.29c}$$

The proportional average saving in number of tests is (from (6.25b))

$$1 - \frac{E}{E[N]} = 1 - p_{\mathrm{OB}} + \frac{(p_{\mathrm{OB}} - p'_{\mathrm{OB}})E[N(1-\omega_{\mathrm{B}})^N] - 1}{E[N]}. \tag{6.30}$$

A-type nonconformity

Direct application of Bayes' theorem leads to

$$\Pr[\text{item is A-type NC}|\text{item is judged A-type NC}]$$
$$= p_{\mathrm{A}}\omega_{\mathrm{A}}\{p_{\mathrm{A}}\omega_{\mathrm{A}} + p'_{\mathrm{A}}(1-\omega_{\mathrm{A}})\}^{-1} = p_{\mathrm{A}}\omega_{\mathrm{A}}\tilde{\omega}_{\mathrm{A}}^{-1}; \tag{6.31a}$$

Pr[item is A-type NC|item is not judged A-type NC]

$$= (1-p_A)\omega_A\{(1-p_A)\omega_A + (1-p'_A)(1-\omega_A)\}^{-1}$$
$$= (1-p_A)\omega_A(1-\tilde{\omega}_A)^{-1}. \tag{6.31b}$$

From Kemp and Kemp's corresponding formula, the proportion of items judged to have NC type A in the accepted output is

$$\tilde{\omega}_A - k(1-\tilde{\omega}_A)/E[N].$$

As a result, the expected proportion of items with NC type A in the accepted output is

$$\frac{p_A\omega_A}{\tilde{\omega}_A}\left\{\tilde{\omega}_A - \frac{k(1-\tilde{\omega}_A)}{E[N]}\right\} + \frac{(1-p_A)\omega_A}{1-\tilde{\omega}}\left\{1-\tilde{\omega} + \frac{k(1-\tilde{\omega}_A)}{E[N]}\right\}$$
$$= \omega_A\left\{1 - \frac{k}{E[N]}\left(\frac{p_A}{\tilde{\omega}_A} - 1\right)\right\}. \tag{6.32}$$

Inspection is of no value unless $p_A > p'_A$, or equivalently $p_A > \tilde{\omega}_A$, in which case the value in (6.32) is less than ω_A.

Tables

Table 1 of Kotz and Johnson (1989), reproduced here as Table 6.3, gives values of the expected length of 'production run' ($E[N]$), and the percentage average saving in inspection for B-type nonconformity for parameter values

$$p = 1{\cdot}0, 0{\cdot}9, 0{\cdot}8;\ p' = 0{\cdot}0, 0{\cdot}1, 0{\cdot}2;\ k = 2, 3, 4;$$
$$\omega_A = 0{\cdot}1, 0{\cdot}2;\ \omega_B = 0{\cdot}001,\ 0{\cdot}0001,$$

with $p_A = p_{OB} = p_B = p$ and $p'_A = p'_{OB} = p'_B = p'$. The table indicates that (i) for large k, (here $k = 4$) $E[N]$ can be too large to be practicable; (ii) taking $k = 2, 3$ gives some protection against 'false positive' ($p'_A > 0$); (iii) when ω_A is small, the 'false negative' ($p_A < 1$) may not be so important.

Considering errors in decisions in regard to A-type nonconformity, the situation is still bearable when the value of p_A is not less than 0·8, but it could be very serious if the value of p'_A is greater than 0·1, especially for smaller k. In regard to B-type nonconformity, errors in inspection are more critical.

Table 6.3a *Values of* $E[N](p_A = p_{0B} = p_B = p;\ p_{AB} = p'_B = p')$

ω_A	k	$p = 1{\cdot}0$			0·9			0·8		
		$p' = 0{\cdot}0$	0·1	0·2	0·0	0·1	0·2	0·0	0·1	0·2
0·1	2	108·00	30·96	14·33	132·57	34·42	15·42	166·75	38·48	16·64
	3	1,107·00	175·76	58·88	1,503·31	204·89	65·23	2,118·88	241·03	72·53
	4	11,106·00	942·09	220·57	16,743·89	1,156·49	252·39	26,531·93	1,437·33	290·36
0·2	2	28·00	14·33	8·49	34·42	16·64	9·59	43·31	19·53	10·89
	3	152·00	58·88	28·93	204·89	72·53	34·03	286·45	90·87	40·41
	4	776·00	220·57	87·46	1,156·49	290·36	107·87	1,811·33	391·27	134·78

Table 6.3b *percentage average savings in inspection for B-type nonconformance* $p_A = p_{0B} = p_B = p$; $p'_A = p'_{0B} = p'_B = p'$

			$p = 1{\cdot}0$			0·9			0·8		
ω_A	k	ω_B	$p' = 0{\cdot}0$	0·1	0·2	0·0	0·1	0·2	0·0	0·1	0·2
0·1	2	0·001	80·88	84·36	79·48	79·61	84·04	78·20	78·27	83·90	77·22
	2	0·0001	97·47	89·42	81·86	97·25	88·97	80·41	97·01	88·66	79·23
	3	0·001	22·43	65·18	72·85	24·29	65·05	72·57	28·17	65·31	72·55
	3	0·0001	81·04	87·32	81·05	77·98	87·03	80·38	74·44	86·82	79·86
	4	0·001	0·67	23·80	54·11	10·28	27·14	54·84	20·10	31·72	56·10
	4	0·0001	22·44	75·29	77·49	22·58	74·34	77·20	25·99	73·54	77·04
0·2	2	0·001	94·86	90·29	86·15	93·66	88·30	82·28	92·68	86·93	79·46
	2	0·0001	100·00	92·97	87·85	99·20	90·07	83·87	98·71	89·60	80·93
	3	0·001	75·61	82·17	80·75	71·99	80·53	78·57	68·21	79·19	77·04
	3	0·0001	97·68	91·39	85·32	96·72	90·27	83·14	95·70	89·46	81·55
	4	0·001	31·66	60·94	69·94	29·29	58·25	68·35	30·07	56·18	67·25
	4	0·0001	86·32	87·23	81·62	82·39	86·17	80·41	77·36	85·15	79·53

CHAPTER 7

Dorfman–Sterrett group testing and its extensions

7.1 Introduction

Some fourteen years after the publication of Dorfman's important note (see Chapter 6), the same journal carried a paper by Andrew Sterrett (Sterrett (1957)), derived from a dissertation submitted to the University of Pittsburgh in June 1956. In this paper Sterrett suggested the following 'anti-Bayesian' modification of Dorfman's procedure.

'Rather than analyse each sample of a defective pool (group) it is proposed to continue making individual tests *only until* a defective is found.' If the proportion of NC items in the population is small it is regarded as likely that a new (sub)group, consisting of the remaining untested units, will prove, on testing, to be free of NC items. If this does happen, the work is finished. Otherwise, individual testing is resumed, until another NC item is found, whereupon the remaining items are tested as a (sub)group, and so on until a decision is reached in regard to each item. The reversion to a group test may be repeated as many times as needed (i.e. up to a possible $(n-2)$ times).

Sterrett calculated that his procedure, when applied continuously (without a stopping rule), increases Dorfman efficiencies on the average by 'about 6%' if optimum group sizes (which are different for Dorfman and Dorfman–Sterrett procedures) are used in each case. Note that the optimal group sizes for Dorfman procedures in Sterrett's Table 1 appear to have been computed from an approximate formula. For example for $\omega = 0{\cdot}35$, the Dorfman–Sterrett group size used is 2. For this group size the two procedures are identical. The Table shows an expected savings of 10% over the Dorfman procedure with 'optimum' group size 3. This cannot be the correct 'optimum' group size, since the procedure with group size 2 (which is also

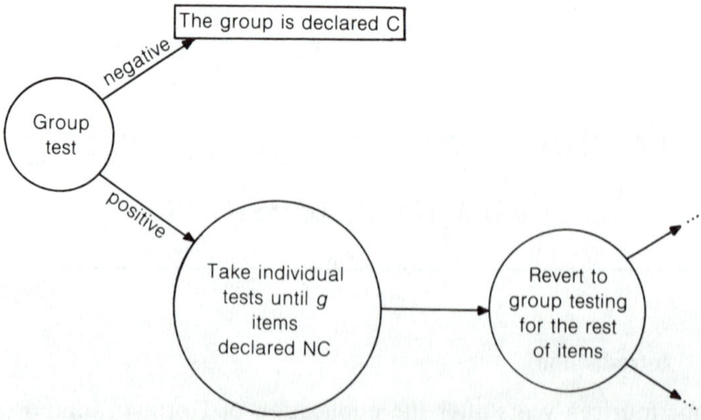

Figure 7.1 *Flow chart for Dorfman–Sterrett procedure* ($g = 1$) *and modified Dorfman–Sterrett procedure* ($g > 1$)

Dorfman – as well as Dorfman–Sterrett) has 10% greater expected savings.

We will discuss restricted Dorfman–Sterrett procedures which limit the number of reversions to a Dorfman procedure. For example, if only one reversion is allowed, we have a one-stage Dorfman–Sterrett procedure. It is clear that the original Sterrett procedure is a $(n-2)$-stage Dorfman–Sterrett procedure.

Modified Dorfman–Sterrett procedures will also be discussed, which are more general than the original one. They use reversion to Dorfman procedures only after $g(\geqslant 2)$ NC decisions, instead of just one. (This is illustrated in Fig. 7.1).

In section 7.4 we will discuss another modification of the Dorfman–Sterrett scheme, wherein after a positive result from the first group testing, we test every item in a subgroup and then either test all items in the next subgroup or test all the remaining subgroups as a single group, depending on the number of NC decisions in the previous subgroup testing. This is termed a **randomized-sequential** procedure.

7.2 Dorfman–Sterrett procedure

Notation

This is similar to Chapter 6. For group inspection, let

$$p_0 = \Pr[\text{identifying an NC group as NC}],$$
$$p'_0 = \Pr[\text{identifying a C group as NC}];$$

and let their counterparts for individual testing be p and p', as before.

Suppose there are D NC items in a lot of size N. It is well known (cf. Chapter 2) that Y, the actual number of NC items in a group of size n has a hypergeometric distribution with parameters $(n; D, N)$.

$$\Pr[Y = y|n] = \frac{\binom{n}{y}\binom{N-n}{D-y}}{\binom{N}{D}} = \frac{\binom{D}{y}\binom{N-D}{n-y}}{\binom{N}{n}}$$

$$\max(0, n-N+D) \leqslant y \leqslant \min(n, D). \quad (7.1)$$

For $N \to \infty$ and $D \to \infty$ with $D/N \to \omega$, or if sampling is with replacement and $D/N = \omega$, we have

$$\Pr[Y = y|n] = \binom{n}{y}\omega^y(1-\omega)^{n-y}, \qquad (0 \leqslant y \leqslant n). \quad (7.2)$$

Here PC(NC), PC(C) and E have the same meanings as in Chapter 6, i.e.

$$\text{PC(NC)} = \Pr(\text{correct identification of an NC item}),$$

$$\text{PC(C)} = \Pr(\text{correct identification of a C item}).$$

$$E = \text{The expected number of tests.}$$

$1 - E/n =$ the expected proportional reduction in testing $\equiv$ EPR. For a k-stage procedure, we will use

$${}^kE, {}^k\text{PC(NC)} \quad \text{and} \quad {}^k\text{PC(C)}$$

to denote the above quantities, emphasizing the number of stages. The original Dorfman procedure corresponds to the $k = 0$ case.

Furthermore we use additional subscripts indicating the group size n and the actual NC number y in the group, on which the corresponding quantities such as ${}^kE$, etc. are conditioned. We therefore have symbols

$${}^kE_n, {}^k\text{PC(NC)}_n, {}^k\text{PC(C)}_n;$$

$${}^kE_{n|y}, {}^k\text{PC(NC)}_{n|y}, {}^k\text{PC(C)}_{n|y}.$$

We will use the following expressions for individual testing:

$$P_{\text{NC}}(m, t|n, y)$$

$$= \begin{cases} \binom{m-1}{t-1}\binom{n-m}{y-t}(1-p)^{t-1}p(1-p')^{m-t}\Big/\binom{n}{y} & (t \geqslant 1) \\ 0 & (t = 0) \end{cases}$$

$$(7.3a)$$

$$P_C(m, t \mid n, y) = \begin{cases} \binom{m-1}{t-1}\binom{n-m}{y-t}(1-p)^t p'(1-p')^{m-t-1} \Big/ \binom{n}{y} & (t \leqslant m-1) \\ 0 & (t = m) \end{cases} \tag{7.3b}$$

where $P_{NC}(m, t \mid n, y)$ (resp. $P_C(m, t \mid n, y)$) is the probability that the first correctly (resp. incorrectly) judged NC item is the mth tested item (events 'A' (resp. 'B')), while in fact there are t truly NC items among the first m tested. The probability that the mth item is the first identified (correctly or incorrectly) as NC is

$$P(m, t \mid n, y) = P_{NC}(m, t \mid n, y) + P_C(m, t \mid n, y). \tag{7.4}$$

This is illustrated in Fig. 7.2. Note that this probability will be zero if m, t, n and y are beyond their natural ranges.

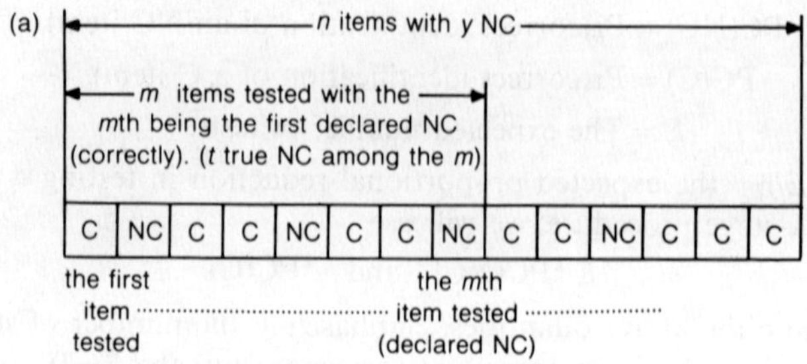

Figure 7.2(a) *For event 'A'*: $\Pr[A] = P_{NC}(m, t \mid n, y); n = 14, m = 8, t = 3, y = 4$

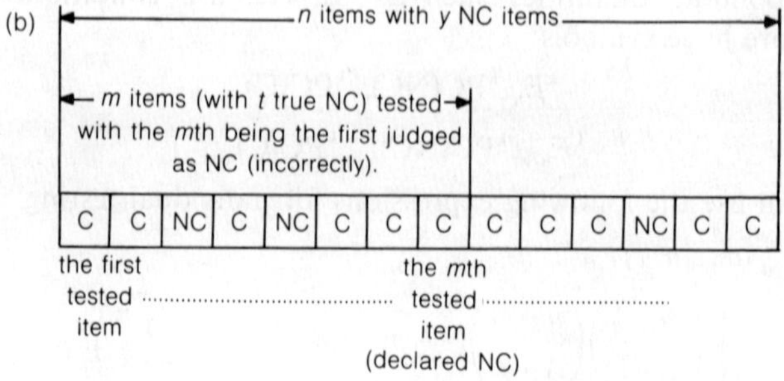

Figure 7.2(b) *For event 'B'*: $\Pr[B] = P_C(m, t \mid n, y); n = 14, m = 8, t = 2, y = 3$

One-stage Dorfman–Sterrett procedures

For $k=0$, the case of simple Dorfman procedures, clearly

$$^{0}E_{n|y}=\begin{cases}1+np_0 & \text{for } y>0\\ 1+np'_0 & \text{for } y=0\end{cases} \tag{7.5}$$

$$^{0}\mathrm{PC(NC)}_{n|y}=p_0p \qquad \text{for all } y \tag{7.6}$$

$$^{0}\mathrm{PC(C)}_{n|y}=\begin{cases}1-p_0p' & \text{for } y>0\\ 1-p'_0p' & \text{for } y=0\end{cases} \tag{7.7}$$

and, taking expectations with respect to Y, we have

$$\begin{aligned}{}^{0}E_n&=\sum_{y=0}^{n}\Pr[Y=y|n]\,{}^{0}E_{n|y}\\ &=1+n\{p_0[1-\Pr[Y=0|n]]+p'_0\Pr[Y=0|n]\},\end{aligned} \tag{7.8}$$

$$^{0}\mathrm{PC(NC)}_n=p_0p, \tag{7.9}$$

$$^{0}\mathrm{PC(C)}_n=1-p'\{p_0(1-\Pr[Y=0|n-1])+p'_0\Pr[Y=0|n-1]\}, \tag{7.10}$$

where $\Pr[Y=0|n-1]$ is the conditional probability that there are no NC items among the other $n-1$ items, given that a specified item in the group of size n is not NC.

In the $k=1$ case, when evaluating ${}^{1}E_n$, since

$$^{1}E_n=\sum_{y=0}^{n}\Pr[Y=y|n]\,{}^{1}E_{n|y}, \tag{7.11}$$

we need to calculate ${}^{1}E_{n|y}$. It is clear that if the mth item is the first identified as NC with t NC items among the m items tested, then there are two possibilities: (1) either $m=n-1$ or $m=n$, in which two cases the total number of tests is always $(n+1)$, with one group test and n individual tests, or (2) for $m\leqslant n-2$, we will start a new Dorfman procedure with group size $n-m$ and $y-t$ NC items in the group after the previous tests of m items. We can therefore write, for $y>0$,

$$\begin{aligned}{}^{1}E_{n|y}=1+p_0\Bigg\{&\sum_{m=1}^{n-2}\sum_{t}^{(y)}P(m,t|n,y)(m+{}^{0}E_{n-m|y-t})\\ &+n\Bigg[1-\sum_{m=1}^{n-2}\sum_{t}^{(y)}P(m,t|n,y)\Bigg]\Bigg\}\end{aligned}$$

$$= 1 + p_0 \left\{ n - \sum_{m=1}^{n-2} \sum_t^{(y)} P(m, t \mid n, y)(n - m - {}^0E_{n-m|y-t}) \right\}$$

$$= 1 + p_0 \left\{ n - \sum_{m=1}^{n-2} \left[\sum_t^{(y-1)} P(m, t \mid n, y)[(n-m)(1-p_0) - 1] \right.\right.$$

$$\left.\left. + P(m, t \mid n, y)[(n-m)(1-p_0') - 1)] \right] \right\}, \qquad (7.12)$$

where the summation $\sum_t^{(z)}$ has the limits over $\max(0, z - n + m) \leqslant t \leqslant \min(m, z)$. In particular, for $y = 0$

$${}^1E_{n|0} = 1 + p_0' \left[\sum_{m=1}^{n-2} P(m, 0 \mid n, 0)(m + {}^0E_{n-m|0}) + n(1-p')^{n-2} \right]$$

$$= 1 + p_0' \left\{ \sum_{m=1}^{n-2} (1-p')^{m-1} p' [m + 1 + (n-m)p_0'] + n(1-p')^{n-2} \right\}$$

$$= 1 + p_0' p'^{-1} \{ 1 + p' - p_0' + np'p_0 - (1-p')^{n-2}(1 - p' - p_0' + 2p'p_0) \}. \qquad (7.13)$$

For evaluating ${}^1\mathrm{PC(NC)}_n$, we have

$${}^1\mathrm{PC(NC)}_n = \sum_{y=1}^{n} P_y^* \, {}^1\mathrm{PC(NC)}_{n|y}, \qquad (7.14)$$

where P_y^* is the probability of $Y = y$ in a group of n, given one specified item is NC, i.e.

$$P_y^* = \begin{cases} \binom{D-1}{y-1}\binom{N-D}{n-y} \Big/ \binom{N-1}{n-1} & \text{for lot size } N; \\ \binom{n-1}{y-1} \omega^{y-1}(1-\omega)^{n-y} & \text{for infinite lot size.} \end{cases} \qquad (7.15)$$

And, noticing that only $y \geqslant 1$ is of interest,

$${}^1\mathrm{PC(NC)}_{n|y} = y^{-1} E[\text{number of NC items correctly classified} \mid Y = y]$$
$$= y^{-1} p_0 E[\text{number of NC items correctly classified} \mid Y = y \text{ and individual testing starts}], \qquad (7.16)$$

where 'individual testing starts' means the group test is positive (NC) with probability p_0.

For $m \leqslant n - 2$, in the subsequent Dorfman procedure for the remaining $y - t$ NC items, the expected number of NC items correctly classified is $(y - t)p_0 p$, where $p_0 p$ is the probability that each item is correctly identified. Therefore,

E[total number of correctly classified NC items]

$$= \begin{cases} P_{\mathrm{NC}}(m,t|n,y) + P(m,t|n,y)(y-t)p_0 p & \text{for } m \leqslant n-2; \\ P_{\mathrm{NC}}(n-1,y-1|n,y) + P(n-1,y-1)|n,y)p & \\ \quad + P_{\mathrm{NC}}(n-1,y|n,y) & \text{for } m = n-1; \\ P_{\mathrm{NC}}(n,y|n,y) & \text{for } m = n. \end{cases} \tag{7.17}$$

Hence, from (7.17), we have

$$\begin{aligned} {}^1\mathrm{PC(NC)}_{n|y} = y^{-1}p_0 \Bigg\{ & \sum_{m=1}^{n-2} \sum_{1}^{(y)} [P_{\mathrm{NC}}(m,t|n,y) + P(m,t|n,y)(y-t)p_0 p] \\ & + P_{\mathrm{NC}}(n-1,y-1|n,y) + P(n-1,y-1|n,y)p \\ & + P_{\mathrm{NC}}(n-1,y|n,y) + P_{\mathrm{NC}}(n,y|n,y) \Bigg\} \qquad (y \geqslant 1). \end{aligned} \tag{7.18}$$

We therefore could evaluate ${}^1\mathrm{PC(NC)}_n$ from (7.14), (7.15) and (7.18).

Similar to (7.14), but for C items, we have

$$ {}^1\mathrm{PC(C)}_n = \sum_{y=0}^{n-1} P_y^{**}\,{}^1\mathrm{PC(C)}_{n|y}, \tag{7.19}$$

where P_y^{**} is the probability of $Y = y$ in a group of n, given one specified item is C, i.e.

$$P_y^{**} = \begin{cases} \dbinom{D}{y}\dbinom{N-D-1}{n-1-y} \Big/ \dbinom{N-1}{n-1} & \text{for lot size } N; \\ \dbinom{n-1}{y}\omega^{y-1}(1-\omega)^{n-1-y} & \text{for infinite lot size.} \end{cases} \tag{7.20}$$

First, for $y > 0$

$$\begin{aligned} {}^1\mathrm{PC(C)}_{n|y} = 1-(n-y)^{-1}p_0 \Bigg[& \sum_{m=1}^{n-2} \Bigg\{ \sum_{1}^{(y-1)} [P_{\mathrm{C}}(m,t|n,y) \\ & + P(m,t|n,y)(n-m-y+t)p_0 p'] \\ & + P_{\mathrm{C}}(m,y|n,y) + P(m,y|n,y)(n-m)p_0' p' \Bigg\} \\ & + P_{\mathrm{C}}(n-1,y|n,y) + P_{\mathrm{C}}(n-1,y-1|n,y) \\ & + P(n-1,y|n,y)p' + P_{\mathrm{C}}(n,y|n,y) \Bigg], \end{aligned} \tag{7.21}$$

and for $y=0$

$$
\begin{aligned}
{}^1\mathrm{PC(C)}_{n|0} \\
&= 1-n^{-1}p_0'\left\{\sum_{m=1}^{n-2}(1-p')^{m-1}p'[1+(n-m)p_0'p']+2p'(1-p')^{n-2}\right\} \\
&= 1-n^{-1}p_0'[1-p_0'+np'p_0'-(1-p')^{n-2}(1-2p')], \qquad (7.22)
\end{aligned}
$$

noting that $(1-p')^{n-2}[p'+p'^2+(1-p')p'] = 2p'(1-p')^{n-2}$.

There is a simpler way of deriving the formula for ${}^1\mathrm{PC(NC)}_n$ for infinite lot size although we could use the formulas just obtained. Consider an NC item, $\mathscr{A}$, say, in the group of size n, with chance $1/n$ to be chosen as the mth tested, with probability p to be classified correctly and with probability $\tilde{\omega}^{m-1}$ to be the first judged NC, where $\tilde{\omega}=\omega p+(1-\omega)p'$ is the probability that a randomly chosen item is declared NC in individual testing. The probability that this particular NC item, $\mathscr{A}$, is the mth tested and is judged NC is:

$$n^{-1}(1-\tilde{\omega})^{m-1}p.$$

Summing over $m=1$ to n, we find

$$
\begin{aligned}
\Pr[\mathscr{A}\text{ is the first item declared NC}] &= n^{-1}p\sum_{m=1}^{n}(1-\tilde{\omega})^{m-1} \\
&= \frac{p\{1-(1-\tilde{\omega})^n\}}{n\tilde{\omega}}. \qquad (7.23)
\end{aligned}
$$

The probability that $\mathscr{A}$ is not among the first m items is $1-m/n$, and that the mth is declared NC is $(1-\tilde{\omega})^{m-1}\tilde{\omega}$. Therefore the probability that the mth tested item is the first declared NC, and $\mathscr{A}$ is among the remaining $n-m$ items is

$$\left(1-\frac{m}{n}\right)(1-\tilde{\omega})^{m-1}\tilde{\omega}.$$

Multiplying this by p_0p (or p for $m=n-1$ only), the probability that the NC item $\mathscr{A}$ is correctly classified as NC, and summing over m, gives

$$
\begin{aligned}
{}^1\mathrm{PC(NC)}_n &= n^{-1}\tilde{\omega}\left[\sum_{m=1}^{n-2}(1-\tilde{\omega})^{m-1}(n-m)p_0p+(1-\tilde{\omega})^{n-2}p\right] \\
&= \frac{p_0p}{n\tilde{\omega}}\{(1-p_0)[1+(2\tilde{\omega}-1)(1-\tilde{\omega})^{n-2}]+p_0n\tilde{\omega}\}, \qquad (7.24)
\end{aligned}
$$

using (7.23).

This approach cannot be used to derive ${}^{1}\mathrm{PC(C)}_n$, because the fact that an NC decision is reached on group test affects the conditional distribution of Y for the group.

Multistage Dorfman–Sterrett procedures

By iteration, using the fact that, after the first reversion to group testing, the key parameters n, y employed before are changed to $n-m$, $y-t$, we can obtain expressions for ${}^{k}E_{n|y}$, ${}^{k}\mathrm{PC(NC)}_{n|y}$ and ${}^{k}\mathrm{PC(C)}_{n|y}$.

For $y > 0$,

$$ {}^{k}E_{n|y} = 1 + p_0 \sum_{m=1}^{n-2} \sum_{t}^{(y)} P(m, t|n, y)(m + {}^{k-1}E_{n-m|y-t}) \tag{7.25a} $$

and

$$ \begin{aligned} {}^{k}E_{n|0} &= 1 + p_0' \sum_{m=1}^{n-2} P(m, 0|n, 0)(m + {}^{k-1}E_{n-m|0}) \\ &= 1 + p_0' \sum_{m=1}^{n-2} (1-p')^{m-1} p'(m + {}^{k-1}E_{n-m|0}). \end{aligned} \tag{7.25b} $$

Also,

$$ \begin{aligned} {}^{k}\mathrm{PC(NC)}_{n|y} = y^{-1} p_0 \Bigg\{ & \sum_{m=1}^{n-2} \sum_{t}^{(y)} [P_{\mathrm{NC}}(m, t|n, y) \\ & + P(m, t|n, y)(y-t)^{k-1}\mathrm{PC(NC)}_{n-m|y-t}] \\ & + P_{\mathrm{NC}}(n-1, y|n, y) + P_{\mathrm{NC}}(n-1, y-1|n, y) \\ & + P(n-1, y-1|n, y)p + P_{\mathrm{NC}}(n, y|n, y) \Bigg\}; \end{aligned} \tag{7.26} $$

and for $y < n$

$$ \begin{aligned} {}^{k}\mathrm{PC(C)}_{n|y} = 1 - (n-y)^{-1} p_0 \Bigg\{ & \sum_{m=1}^{n-2} \sum_{t}^{(y)} [P_{\mathrm{C}}(m, t|n, y) \\ & + P(m, t|n, y)(n-m-y+t)(1 - {}^{k-1}\mathrm{PC(C)}_{n-m|y-t})] \\ & + P_{\mathrm{C}}(n-1, y|n, y) + P(n-1, y|n, y)p' \\ & + P_{\mathrm{C}}(n-1, y-1|n, y) + P_{\mathrm{C}}(n, y|n, y) \Bigg\}. \end{aligned} \tag{7.27a} $$

For $y = 0$,

$$ {}^{k}\mathrm{PC(C)}_{n|0} = 1 - n^{-1}p'_0\left\{\sum_{m=1}^{n-2}(1-p')^{m-1}p'[1+(n-m)^{k-1}\mathrm{PC(C)}_{n-m|0}] + 2p'(1-p')^{n-2}\right\}. \quad (7.27b) $$

Tables

Tables 1–6 of Johnson, Kotz and Rodriguez (1990) give values of ${}^{k}E_{n|y}$, ${}^{k}\mathrm{PC(NC)}_{n|y}$ and ${}^{k}\mathrm{PC(C)}_{n|y}$ for various parameter values:

$$ k = 1,2; \quad n = 6; \quad p_0 = p = 0{\cdot}75, 0{\cdot}90, 0{\cdot}95; $$
$$ p'_0 = p' = 0{\cdot}05, 0{\cdot}10, 0{\cdot}25; \quad y = 0, 1, \ldots, 6. $$

Values of ${}^{k}E_{n|y}$ change little when $y = 0$ but more when y is larger, especially for higher values of p_0 and p. When ω or D/N is small, the original Dorfman procedure might be preferable, but when ω or D/N is greater, one can see the advantage of the multistage Dorfman–Sterrett procedure. With increasing k, ${}^{k}\mathrm{PC(C)}_{n|y}$ increases, while ${}^{k}\mathrm{PC(NC)}_{n|y}$ decreases slightly, due to the effect of group testing.

Tables 7–9 of Johnson, Kotz and Rodriguez (1990) give comparisons of the three quantities between cases $k = 1$ and $k = 2$ for different values of y.

7.3 Modified Dorfman–Sterrett screening procedures

Introduction

In this section, we will study modified Dorfman–Sterrett procedures, in which, during the subsequent individual testing after a positive result from a group test, reversion to Dorfman screening is not undertaken until the gth NC declaration ($g \geqslant 2$) occurs. The original Dorfman–Sterrett procedure is equivalent to the case $g = 1$. This modification could be more effective than Sterrett's when ω is somewhat higher, (though still quite low). Because more NC items have been removed before reversion to Dorfman screening, it is more likely that the later group testing will give a negative result, and further individual testing will not be necessary.

We will restrict our analysis to one-stage procedures.

Notation

The notation is the same as in previous sections except that we will omit the superscript '1', in quantities such as ${}^1E_{n|y}$, ${}^1\mathrm{PC(NC)}_{n|y}$ and ${}^1E_{n|y}$ conditioned on $Y=y$, the number of NC items in a group of size n. We could evaluate the corresponding overall values by compounding them with the appropriate distribution of Y. We will use the relations

$$\mathrm{PC(C)}_{n|y} = E(\mathrm{C})_{n|y}/(n-y); \qquad \mathrm{PC(NC)}_{n|y} = E(\mathrm{NC})_{n|y}/y, \quad (7.28)$$

where $E(\mathrm{C})_{n|y}$, $E(\mathrm{NC})_{n|y}$ denote the expected number of correctly identified C and NC items, respectively. Symbols p, p', p_0 and p'_0 will have the same meanings as in the previous section.

Let M be the number of items tested such that the Mth item is the gth item declared NC, and let T be the number of truly NC items among those M tested items, and let J be the number of truly C items among those declared NC. In addition to situations conditioned on $Y=y$, we also consider conditioning on $M=m$, $T=t$ and $J=j$. We will use the set of symbols $\{n|m,t,y,j\}$ for that purpose.

We define

$$P(m,t,j|n,y) \equiv \Pr[(M=m)\cap(T=t)\cap(J=j)|\,Y=y]. \quad (7.29)$$

Also

$$P(m,t|n,y) = \sum_j P(m,t,j|n,y) \equiv \Pr[(M=m)\cap(T=t)|\,Y=y], \quad (7.30)$$

with $\max(0, g-t) \leqslant j \leqslant \min(g, m-t)$. (This is illustrated in Fig. 7.3.)

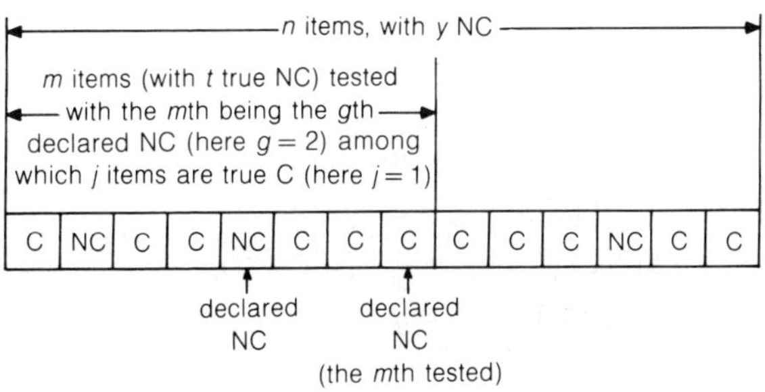

Figure 7.3 *For event 'A'*: $\Pr[\mathrm{A}] = P(m,t,j|n,y)$; $n=14, m=8, t=2, y=3, j=1, g=2$

It is assumed that the results of any tests (group or individual) are mutually independent.

Analysis

For $y>0$

$$\begin{aligned}P(m,t,j\,|\,n,y) = {} & \Pr[\text{group test gives an NC result}]\\ & \times \Pr[(t\text{ NC's and }(m-t)\text{ C's in first }n\text{ items tested})\\ & \cap((g-j)\text{ NC's and }j\text{ C's declared NC})]\\ & \times \Pr[g\text{th item declared NC is }m\text{th tested}]\\ = {} & p_0\binom{n}{y}^{-1}\binom{n-m}{y-t}\binom{m}{t}\binom{t}{g-j}\binom{m-t}{j}\\ & \times p^{g-j}(1-p)^{t-g+j}p'^{j}(1-p')^{m-t-j}\times gm^{-1},\end{aligned}\tag{7.31}$$

where $\max(0, y-n+m) \leqslant t \leqslant \min(y, m)$; $\max(0, g-t) \leqslant j \leqslant \min(g, m-t)$.

If $y=0$, p_0 is replaced by p_0' and only $j=g$ is relevant. If $y=n$, only $j=0$ is relevant.

For $2 \leqslant m \leqslant n-2$

$$E(\mathrm{C})_{n|m,t,y,j} = m-t-j+(n-y-m+t)\{1-h(t,y)p'\},$$
$$E(\mathrm{NC})_{n|m,t,y,j} = g-j+(y-t)p_0p,$$
$$E_{n|m,t,y} = m+2+(n-m)h(t,y),$$

where

$$h(t,y)=\begin{cases}p_0 & \text{if } t<y;\\ p_0' & \text{if } t=y.\end{cases}$$

Note that $E_{n|m,t,y,j} = E_{n|m,t,y}$, not depending on j.

Also

$$E(\mathrm{C})_{n|n-1,t,y,j} = n-y-j-(t+1-y)p'.$$
$$E(\mathrm{C})_{n|n,y,y,j} = n-y-j,$$
$$E(\mathrm{NC})_{n|n-1,t,y,j} = g-j+(y-t)p,$$
$$E(\mathrm{NC})_{n|n,y,y,j} = g-j,$$

and

$$E_{n|n-1,y,y,j} = E_{n|n,y,y} = n+1.$$

In the following evaluation, we should keep in mind that it is possible to have fewer than g items declared NC, or to have a negative result on group testing, so that all items in the group are declared C.

We have

$$E(\mathrm{C})_{n|y} = \sum_{m=g}^{n} \sum_{t} \sum_{j} P(m,t,j|n,y)E(\mathrm{C})_{n|m,t,y,j}$$

$$+ p_0 \sum_{h=0}^{g-1} \sum_{j=0}^{h} (n-y-j)\binom{y}{h-j}\binom{n-y}{j} p^{h-j}(1-p)^{y-h+j}p'^{j}$$

$$\times (1-p')^{n-y-j} + (n-y)\{1-h(n-y,n)\}, \tag{7.32}$$

with limits $\max(0, y-n+m) \leqslant t \leqslant \min(y,m)$;
$\max(0, g-t) \leqslant j \leqslant \min(g, m-t)$.

$$E(\mathrm{NC})_{n|y} = \sum_{m=g}^{n} \sum_{t} \sum_{j} P(m,t,j|n,y)E(\mathrm{NC})_{n|m,t,y,j}$$

$$+ p_0 \sum_{h=0}^{g-1} \sum_{j=0}^{h} (h-j)\binom{y}{h-j}\binom{n-y}{j} p^{h-j}$$

$$\times (1-p)^{y-h+j}p'^{j}(1-p')^{n-y-j}, \tag{7.33}$$

with limits $\max(0, y-n+m) \leqslant t \leqslant \min(y,m)$;
$\max(0, g-t) \leqslant j \leqslant \min(g, m-t)$.

For $y > 0$, we have the following formulas for expected numbers of tests:

$$E_{n|y} = \sum_{m=g}^{n} \sum_{t} P(m,t|n,y)E_{n|m,t,y} + np_0 \sum_{h=0}^{g-1} \sum_{j=0}^{h} \binom{y}{h-j}\binom{n-2-y}{j}$$

$$\times p^{h-j}(1-p)^{y-h+j}p'^{j}(1-p')^{n-2-y-j} + 1 - p_0. \tag{7.34a}$$

For $y = 0$,

$$E_{n|0} = 1 + p_0'\left\{\sum_{m=g}^{n-2}\binom{m-1}{g-1}p'^{g}(1-p')^{m-g}\{m+1+(n-m)p_0'\}\right.$$

$$\left. + n\sum_{j=0}^{g-1}\binom{n-2}{j}p'^{j}(1-p')^{n-2-j}\right\}, \tag{7.34b}$$

with limits for t: $\max(0, y-n+m) \leqslant t \leqslant \min(y,m)$.

Tables

Values of

$$E_{n|y}, \qquad \mathrm{PC(NC)}_{n|y} = E(\mathrm{NC})_{n|y}, \qquad \mathrm{PC(C)}_{n|y} = E(\mathrm{C})_{n|y}/(n-y),$$

for all relevant values of y are given in Tables 1, 2 and 3 of Huang *et al.* (1989) for the case $g=2$; $n=6,8$, respectively. Similar, comparable, tables of

$$^kE_{n|y}, \qquad ^k\text{PC(NC)}_{n|y}, \qquad ^k\text{PC(C)}_{n|y}$$

are given by Johnson *et al.* (1987), for $k=1,2$; $n=6$, for standard Dorfman–Sterrett procedures (cf. section 7.2).

Comparing tables, although values of $E_{n|y}$ for the modified procedure are higher, values of PC(NC) are substantially greater.

Because the modified procedure uses more group tests, the chances of a C decision increase. The progression of values of $E_{6|y}$ and $\text{PC(C)}_{6|y}$ with respect to y is rather irregular, and $\text{PC(C)} < {}^1\text{PC(C)} < {}^2\text{PC(C)}$, but $\text{PC(NC)} > {}^1\text{PC(NC)} > {}^2\text{PC(NC)}$.

A special case

We can obtain PC(NC) directly, without first evaluating $\text{PC(NC)}_{n|y}$, when the lot size is large (effectively infinite). Denote the proportion of NC items in the population by ω. The probability that a randomly chosen item is declared NC in individual testing is denoted by $\tilde{\omega} = \omega p + (1-\omega)p'$, as before. Let $\mathscr{A}$ be a specific NC item. Let A denote the event that in individual testing, there is a correct NC decision for $\mathscr{A}$, and there are h NC decisions among the other $n-1$ items. We have

$$P_h \equiv \Pr[A] = p\binom{n-1}{h}\tilde{\omega}^h(1-\omega)^{n-1-h}.$$

Given event A, the conditional probability that there will be an NC decision for $\mathscr{A}$ is

$$\Pr[\mathscr{A} \text{ is declared NC}\,|\,A]$$

$$= \begin{cases} 1 & \text{if } h<g; \\ \dfrac{g}{g+1} + \dfrac{g}{n(n-1)} + \left\{1 - \dfrac{g}{g+1} - \dfrac{g}{n(n+1)}\right\}p_0 & \text{if } h=g; \\ \dfrac{g}{h+1} + \left(1 - \dfrac{g}{h+1}\right)p_0 & \text{if } h>g. \end{cases} \tag{7.35}$$

Note that conditioned on A, (i) when $h<g$, individual testing will continue until either the nth item is tested (if $h<g-1$) or the $(h+1)$th

item is declared NC (if $h+1=g$) so that $\mathscr{A}$ is among those $h+1$ items declared NC, and there is no need for group testing; (ii) when $h>g$, either $\mathscr{A}$ is among the first g items declared NC with probability $g/(h+1)$, or there is a further group test before $\mathscr{A}$ is detected; (iii) if $g=h$, no second group test is needed if either $\mathscr{A}$ is among the first g items declared NC as in (ii), with probability $g/(h+1)=g/(g+1)$, or after g NC decisions all items in the group have been tested except one, which is $\mathscr{A}$, with probability $g/(n(n-1))$; otherwise there would be a second group test before $\mathscr{A}$ is detected. Hence

$$\begin{aligned}\text{PC(NC)}\\ &= p_0\left[\sum_{h=0}^{g-1} P_h + \left\{\frac{g}{g+1}+\frac{g}{n(n-1)}+\left[1-\frac{g}{g+1}-\frac{g}{n(n-1)}\right]p_0\right\}P_g\right.\\ &\quad \left.+\sum_{h=g+1}^{n-1}\left\{\frac{g}{h+1}+\left[1-\frac{g}{h+1}\right]p_0\right\}P_h\right]. \qquad (7.36)\end{aligned}$$

The idea of this method is that, given $\mathscr{A}$, the positive result of the group testing (i.e. starting individual testing) has probability p_0, whatever may be the status of the remaining $n-1$ items. However this idea does not work for evaluation of PC(C), since our procedure depends on the number of NC decisions (cf. pp. 104–5).

7.4 Randomized-sequential group testing procedures

Introduction

In a Dorfman–Sterrett procedure, (a) if a group (of size n) test gives a negative result, we declare all items C in the group, while for a positive group result, (b) we continue with individual testing and take a reversion to group testing if a NC item is declared. Now we modify the (b) part of the procedure. We divide the group into h subsets of size n_1 each. If step (a) gives a positive result, which indicates the presence of at least one NC item in the group of n, we test all n_1 items in the first subset individually, and, depending on the number Z_1 of NC decisions in the first subset, we *either* test every item individually in the second subset (taking us back to the same situation as testing the first), *or* apply group testing to the remaining $n-n_1$ items (which takes us back to (a) with $n-n_1$ replacing n). We could continue this procedure. Generally, after testing the ith subset, since the number of NC decisions Z_i is random, the decision whether to take individual testing in subset $(i+1)$ or apply a group test to the remaining $n - in_1$ items is a sequential one. It is clear that the special case with $n_1 = 1$ and $h = n$ is the original Dorfman–Sterrett procedure where subsets are individual items and Z_i's are either 1, leading to group testing, or 0, leading to continuation of individual testing.

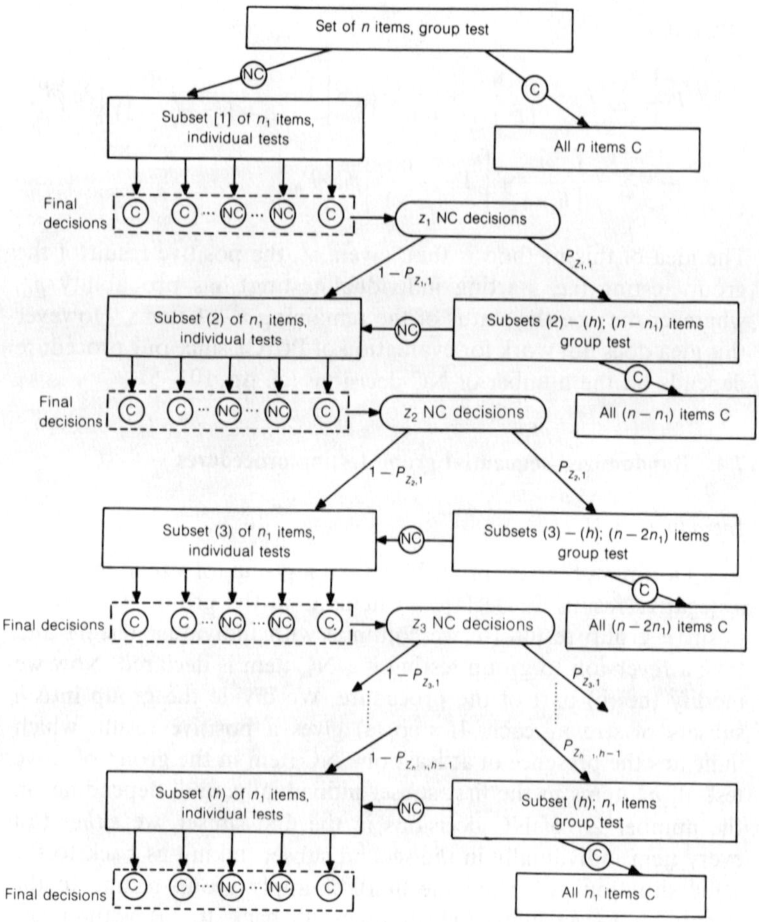

Figure 7.4 *Random-sequential procedure – flow chart*

Note that we do not form any further subgroups from the existing h subsets of size n_1. Figure 7.4 gives a diagrammatic representation of the procedure.

Notation and assumptions

We assume the probabilities of decisions for any group test are independent of the size of the group, and the results of tests are mutually independent. We can then define p (resp. p_0) as the probability of a correct individual (resp. group) NC decision, and p' (resp. p'_0) as the probability of an incorrect individual (resp. group) NC decision. Let Y_j be the number of truly NC items in the jth subset of size n_1, and J_0 be that in the original set of size n ($hn_1 = n$). Conditioned on $Y_j = y_j$, Z_j, the number of NC decisions in individual testing of the jth subset, is the sum of two independent binomial variables with parameters (y_j, p) and $(n_1 - y_j, p')$, respectively, for $j = 1, \ldots, h$. The conditional distribution of Z_j is

$$Z_j | Y_j = y_j \sim \text{Bin}(y_j, p) * \text{Bin}(n_1 - y_j, p'), \tag{7.37a}$$

or

$$\pi_{z_j|y_j} \equiv \Pr[Z_j = z_j | y_j]$$
$$= \sum_{i=0}^{z_j} \binom{y_j}{i} \binom{n_1 - y_j}{z_j - i} p^i p'^{z_j - i} (1-p)^{y_j - i} (1-p')^{n_1 - y_i - z_j + i}, \tag{7.37b}$$

with $z_j = 0, \ldots, n_1$. Let $P_{z_j, j}$ be the probability that, given $Z_j = z_j$, the remaining $(n - hn_1)$ items are subject to a group test instead of individual testing for subset $(j+1)$, conditioned on Y_j. Then the probability of resorting to group testing is

$$\gamma_j \equiv \gamma_{j|y_j} \equiv \sum_{z_j=0}^{n_1} \pi_{z_j|y_j} P_{z_j, j}. \tag{7.38}$$

The values $\{P_{z,j}\}$ constitute the **inspection strategy**. The original Dorfman procedure corresponds to the case that $P_{z_j, j} = 0$ for all j. So far as possible, we try to choose inspection strategies leading to good values for E, PC(NC) and PC(C).

Assume the sample of size n has been taken from a large population, effectively infinite, and therefore the proportion of NC items is a constant denoted by ω.

Let $\mathbf{Y} \equiv (Y_1, \ldots, Y_h)$, $\mathbf{y} \equiv (y_1, \ldots, y_h)$, $t_j \equiv y_{j+1} + \cdots + y_h$ (Note:

$t_0 = y_1 + \cdots + y_h \leqslant n$) denote the number of NC items in the rest of the $(h-j)$ subsets, and define

$$\phi_j \equiv \begin{cases} p_0 & \text{if } t_j > 0 \\ p'_0 & \text{if } t_j = 0 \end{cases} \tag{7.39}$$

with $j = 0, 1, \ldots, h-1$, which is the probability of reaching an NC decision for the group formed by the union of the $(h-j)$ subsets with a total of t_j NC items.

Analysis

Given that the jth subset is tested individually, the $(j+1)$th subset is also tested individually if and only if the remaining $(h-j)$ subsets are not tested as one group (with probability $1-\gamma_j$) or they are tested as one group with a positive result (with probability $\gamma_j\phi_j$). Therefore conditional on the jth subset being tested individually, the probability that the $(j+1)$th subset is also tested individually is

$$\theta_j \equiv \begin{cases} 1 - \gamma_j + \gamma_j\phi_j & \text{if } j = 1, \ldots, h-1; \\ \phi_0 & \text{if } j = 0. \end{cases} \tag{7.40}$$

Note that θ_0 is the probability that the first subset is tested individually. From (7.40), the probability that the jth subset is tested individually is $\prod_{i=0}^{j-1}\theta_i$, $(j = 2, \ldots, h)$. For each $j = 1, \ldots, h$, when the jth subset is tested individually, there will certainly be n_1 individual tests, plus one group test, of the union of the remaining $(h-j)$ subsets with probability γ_j. Hence the expected number of tests, given $\mathbf{Y} = \mathbf{y}$ is:

$$E_{\mathbf{y}} = 1 + \sum_{j=1}^{h} \left\{ \prod_{i=1}^{j-1} \theta_i \right\} (n_1 + \gamma_j) \qquad \text{with } \gamma_h = 0. \tag{7.41}$$

Conditional on $\mathbf{Y} = \mathbf{y}$, the probability of correctly identifying any given NC item is:

$$\begin{aligned} \mathrm{PC}(\mathrm{NC}\,|\,\mathbf{y}) &= \sum_{j=1}^{h} \Pr[\text{NC is in } j\text{th subset}] \\ &\qquad \times \Pr[j\text{th subject is tested individually}]p \\ &= \sum_{j=1}^{h} \left\{ (y_j t_0^{-1}) \left(\prod_{i=0}^{j-1} \theta_i \right) \right\} p = (p t_0^{-1}) \sum_{j=1}^{h} \left(y_j \prod_{i=0}^{j-1} \theta_i \right). \end{aligned} \tag{7.42}$$

Similarly, for the probability of correctly identifying a given C item given $\mathbf{Y}=\mathbf{y}$, we have

$$\begin{aligned}
\mathrm{PC}(\mathrm{C}\mid\mathbf{y}) &= \sum_{j=1}^{h} \Pr[\mathrm{C} \text{ is in } j\text{th subset}] \\
&\quad \times \{\Pr[j\text{th subset not tested individually}] \\
&\quad + \Pr[j\text{th subset tested individually}](1-p')\} \\
&= \sum_{j=1}^{h} \left\{ \frac{n_1-y_j}{n-t_0} \left[1 - \prod_{i=0}^{j-1} \theta_i + \left(\prod_{i=0}^{j-1} \theta_i \right)(1-p') \right] \right\} \\
&= 1 - p'(n-t_0)^{-1} \sum_{j=1}^{h} \left\{ (n_1-y_j) \left(\prod_{i=0}^{j-1} \theta_i \right) \right\}. \qquad (7.43)
\end{aligned}$$

We will now obtain the three quantities, unconditioned, by averaging (7.41), (7.42) and (7.43) over $\mathbf{y}$. First, recall

$$\begin{aligned}
\Pr[\mathbf{Y}=\mathbf{y}] = P_{\mathbf{y}} &= \prod_{j=1}^{h} \binom{n_1}{y_j} \omega^{y_j}(1-\omega)^{n_1-y_j} \\
&= \left\{ \prod_{j=1}^{h} \binom{n_1}{y_j} \right\} \omega^{t_0}(1-\omega)^{n-t_0}. \qquad (7.44)
\end{aligned}$$

The expected number of tests is

$$E = \sum_{\mathbf{y}} P_{\mathbf{y}} E_{\mathbf{y}}, \qquad (7.45)$$

and the expected percentage reduction as compared with simple individual inspection, in the number of tests is

$$100\,\mathrm{EPR} = 100(1 - E/n).$$

Since $\mathrm{PC}(\mathrm{NC}\mid\mathbf{y})$ applies to each of t_0 NC items, the probability of correct indication of an NC item is

$$\begin{aligned}
\mathrm{PC}(\mathrm{NC}) &= \sum_{\mathbf{y}} t_0 P_{\mathbf{y}} \mathrm{PC}(\mathrm{NC}\mid\mathbf{y}) \Big/ \left(\sum_{\mathbf{y}} t_0 P_{\mathbf{y}} \right) \\
&= \sum_{\mathbf{y}} t_0 P_{\mathbf{y}}\, \mathrm{PC}(\mathrm{NC}\mid\mathbf{y})/E[T_0] \\
&= \sum_{\mathbf{y}} t_0 P_{\mathbf{y}}\, \mathrm{PC}(\mathrm{NC}\mid\mathbf{y})/(n\omega). \qquad (7.46)
\end{aligned}$$

Similarly, for a C item, we have

$$\begin{aligned}\text{PC(C)} &= \sum_{\mathbf{y}} (n-t_0)P_{\mathbf{y}}\text{PC}(\text{C}|\mathbf{y}) \Big/ \left(\sum_{\mathbf{y}} (n-t_0)P_{\mathbf{y}}\right) \\ &= \sum_{\mathbf{y}} (n-t_0)P_{\mathbf{y}}\,\text{PC}(\text{C}|\mathbf{y})/\{n(1-\omega)\}. \qquad (7.47)\end{aligned}$$

Tables and comments

Before studying specific examples, one can see that

$$\text{(i) PC(NC)} \leqslant p_0 p \qquad \text{and} \qquad \text{(ii) PC(C)} \geqslant 1 - p'.$$

Note that an item can be declared NC only by individual test while an item can be declared C by either individual or group testing. We would expect, therefore, that any modification that increases the amount of group testing is likely to increase PC(C), but decrease PC(NC).

Tables I and II of Johnson *et al.* (1990) provide values of EPR, PC(C) and PC(NC) for $n = 12$, with different sets of values of parameters n_1, p_0, p, p' and $P_{z,j}$. These tables reveal that EPR is not greatly affected by inspection strategy but increases as ω decreases, and PC(NC) is affected more seriously by inspection strategy.

If we want to maximize only one of the values PC(C) and PC(NC), we could take a deterministic strategy, with each $P_{z,j}$ equal to either 0 or 1, which will be nearly optimal. But if we need some compromise between PC(C) and PC(NC), a probabilistic strategy may be suitable.

We now present a few specific cases. We take $n = 12$, $n_1 = 2$ (Table I) and inspection strategy

$$P_{0,j} = 0; \qquad P_{1,j} = 1 - 0{\cdot}5^j; \qquad P_{2,j} = 1 \qquad (j = 1, \ldots, h-1)$$

with two sets of decision probabilities

1. $p_0 = p = 0{\cdot}98$; $p'_0 = p' = 0{\cdot}05$;
2. $p_0 = 0{\cdot}9$; $p = 0{\cdot}95$; $p'_0 = p' = 0{\cdot}05$.

Comparing corresponding values of EPR, PC(C) and PC(NC) in Table I with those from simple Dorfman procedures (see Johnson *et al.* (1990)), we find that the random-sequential procedure substantially reduces EPR. and increases PC(C), but also decreases PC(NC), which might be very important if the nonconformity is of a serious nature. For either procedure, values of PC(NC) will increase with increasing p_0 and p, but large values of p_0 and p will prevent PC(NC) from declining drastically, whatever procedure is used.

CHAPTER 8

Curtailed Dorfman-type procedures

8.1 Introduction and notation

In previous chapters, we have discussed a variety of group screening procedures. All of them have one thing in common: that the group testing can only indicate the existence of one or more NC items. Motivated by Pfeifer and Enis (1978), we now study situations in which the group testing indicates not only the presence but also the number of NC items in a group. In the errorless case, Pfeifer and Enis (1978) devised methods of using the extra knowledge about the number of NC items in the group to reduce the expected number of tests. The improvement due to the additional information can be illustrated by the following example. Suppose a test performed on a group of four items provides information that three NC items and one C item are present. If the first item tested individually is identified as C, the remaining items are then automatically identified (as NC) without any further testing. Note that if the additional information were not available, then four tests would be required even using a Dorfman-type procedure (except that if the first three individual tests gave C results, we would know, without testing, that the remaining item 'must' be NC, since the group test indicated the presence of at least one NC item).

More generally, let Y be the number of truly NC items in a group of size n_0. If the inspection is perfect, then the group test estimates the value of Y. If $Y = 0$, then all items are C, and no further testing is needed. If $Y \geqslant 1$, then individual testing continues until either Y NC decisions have been obtained (the remaining, untested, items 'must' be all C) or $n_0 - Y$ not NC decisions have been obtained (the untested items 'must' be NC), whichever comes first. This procedure is called **curtailed** group testing.

We now discuss the effects of errors in inspection on these curtailed group testing procedures and in particular the savings in the expected number of tests. As before, p and p' denote the probabilities of correctly and incorrectly indicating an NC item on individual inspection, respectively; p_1 and p'_1 refer to the similar probabilities contributing to the estimated number of NC items from the group test, for each item. We also use Z_1 to denote the number of declared NC items as estimated from the group test and Z_2 for the number of *potentially* declared NC items if all the items were tested individually; and let M be the number of individual tests carried out.

Acting on the assumption that inspection is error-free, we would first apply the group test and get Z_1 NC items among the n_0 items. If $Z_1 = 0$, we declare all n_0 items C. If $Z_1 > 0$, as the second step, we take individual tests until either Z_1 NC items are identified (the rest are declared as C) or $n_0 - Z_1$ C items are identified (the rest are declared as NC) whichever comes first. (This is illustrated in Figs. 8.1a, 8.1b and 8.2.)

It is easy to see that, given Y, Z_1 (as well as Z_2) is distributed as

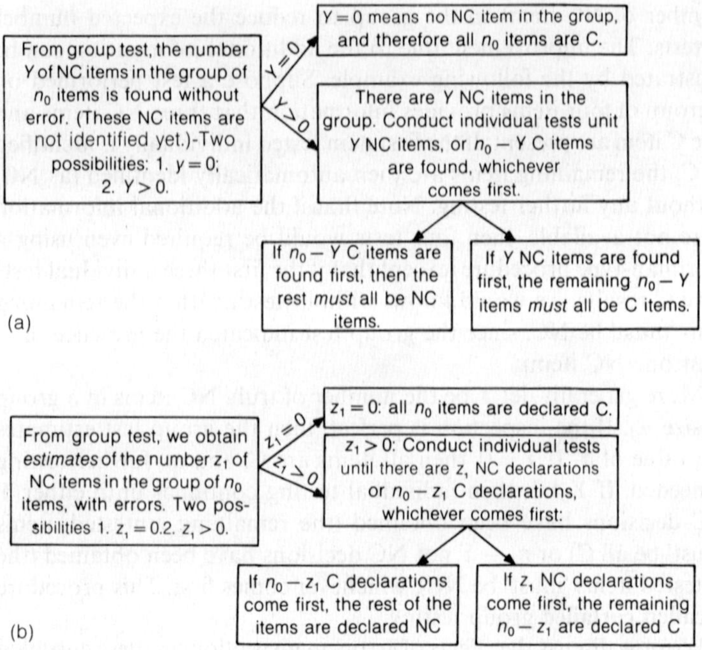

Figure 8.1 *Curtailed group testing (a) without error, and (b) with errors.*

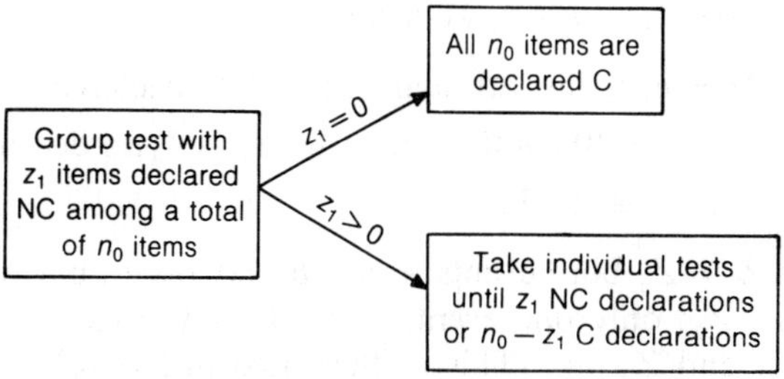

Figure 8.2 *Simplified flow chart for curtailed group testing procedure*

the convolution of two binomial distributions (sum of numbers of items correctly and incorrectly classified as NC), i.e.

$$Z_1 | Y \sim \text{Bin}(Y, p_1) * \text{Bin}(n_0 - Y, p'_1);$$
$$Z_2 | Y \sim \text{Bin}(Y, p) * \text{Bin}(n_0 - Y, p'),$$

whence

$$\Pr[Z_1 = z_1 | n_0, y, p_1, p'_1] = P(z_1 | n_0, y, p_1, p'_1), \tag{8.1}$$

$$\Pr[Z_2 = z_2 | n_0, y, p, p'] = P(z_2 | n_0, y, p, p'), \tag{8.2}$$

where

$$P(z | n_0, y, a, b) \equiv \sum_{w=0}^{z} \binom{y}{w} \binom{n_0 - y}{z - w} a^w b^{z-w} (1-a)^{y-w} (1-b)^{n_0 - y - z + w}. \tag{8.3}$$

We will assume Z_1 and Z_2 to be mutually independent. We will also use the relation

$$\sum_{a=c_1}^{n-c_2} \binom{a}{c_1} \binom{n-a}{c_2} = \binom{n+1}{c_1 + c_2 + 1},$$

with $\binom{a}{b} = 0$ if $a < 0$, $b < 0$ or $b > a$.

8.2 Curtailed group testing

Distribution of number of tests

By the description of our procedure in Section 8.1,

$$(M = 0) \equiv (Z_1 = 0) \cup (Z_1 = n_0)$$

and for $m = 1, \ldots, n-1$, we have

$$\begin{aligned}(M = m) &\equiv (\text{the } m\text{th item is the } Z_1\text{th declared NC}) \\ &\quad \cup (\text{the } m\text{th item is the } (n_0 - Z_1)\text{th declared C}) \\ &\equiv A \cup B.\end{aligned}$$

When $Z_2 = Z_1$ both events A and B may occur, but when $Z_2 > Z_1$ or $Z_2 < Z_1$, only one event, A or B, may occur, conditioned on $Z_1 = z_1$ and $Z_2 = z_2$ (This is illustrated in Fig. 8.3). Therefore, for $m = 1, 2, \ldots, n-1$, we have the conditional probabilities

$$\Pr[M = m | z_1, z_2] = \begin{cases} \binom{m-1}{z_1 - 1}\binom{n_0 - m}{z_2 - z_1} \Big/ \binom{n_0}{z_2} & \text{if } z_2 > z_1; \\ \left\{\binom{m-1}{z_1 - 1} + \binom{m-1}{n_0 - z_1 - 1}\right\} \Big/ \binom{n_0}{z_1} & \text{if } z_2 = z_1; \\ \binom{m-1}{n_0 - z_1 - 1}\binom{n_0 - m}{z_1 - z_2} \Big/ \binom{m}{z_2} & \text{if } z_2 < z_1. \end{cases} \tag{8.4}$$

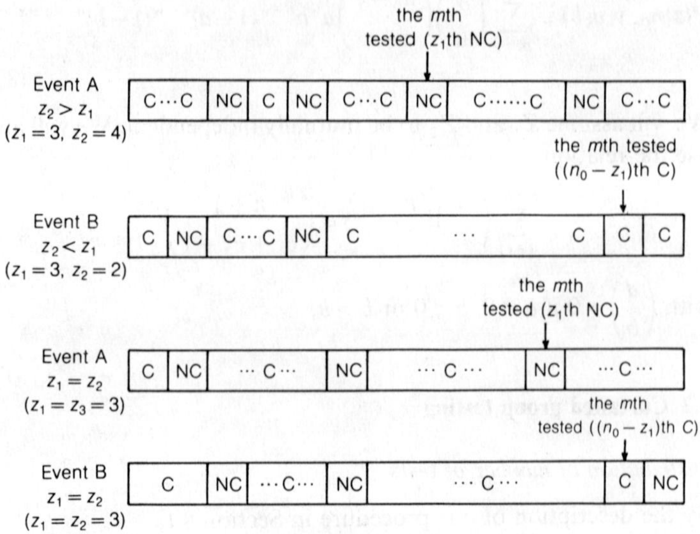

Figure 8.3 *Examples of curtailed group testing procedure*

Also, by definition of M,

$$\Pr[M=0|0,z_2]=\Pr[M=0|n_0,z_2]=1 \qquad \text{for all } z_2 \tag{8.5}$$

and

$$\Pr[M=0|z_1,z_2]=0 \qquad \text{for all } z_1 \neq 0, n. \tag{8.6}$$

Using (8.4) and the identity at the end of section 8.1, direct calculation yields the conditional expected value of M as:

$$E[M|z_1,z_2]=\begin{cases}(n_0+1)z_1/(z_2+1) & \text{if } z_2>z_1\\ z(n_0-z)\{(z+1)^{-1}+(n_0-z+1)^{-1}\} & \text{if } z_2=z_1=z\\ (n_0+1)(n_0-z_1)/(n_0-z_2+1) & \text{if } z_2<z_1.\end{cases} \tag{8.7}$$

The expected total number of tests is $1+E[M]$, where

$$\begin{aligned}E[M]&=E_Y[E_Z(E[M|Z_1,Z_2])|Y]\\ &=\sum_y P_y\sum_{z_1=0}^{n_0}\sum_{z_2=0}^{n_0}P(z_1|n_0,y,p_1,p_1')P(z_2|n_0,y,p,p')E[M|z_1,z_2],\end{aligned} \tag{8.8}$$

where, for lot size N with D NC items,

$$P_y\equiv\Pr[Y=y]=\binom{D}{y}\binom{N-D}{n_0-y}\Big/\binom{N}{n_0},$$

$$\text{if} \quad \max(0,D+n_0-N)\leqslant y\leqslant\min(n_0,D), \tag{8.9a}$$

and for infinite lot size,

$$P_y\equiv\Pr[Y=y]=\binom{n_0}{y}\omega^y(1-\omega)^{n_0-y}, \qquad \text{if } 0\leqslant y\leqslant n_0 \tag{8.9b}$$

where ω is the proportion of NC items.

Values of $E[M]$ calculated from (8.8) for a number of sets of parameter values are given in Tables 9.1 and 10.1 of Johnson *et al.* (1988) and are partially reproduced in section 8.3.

Evaluation of PC(NC)

First, we specify an NC item, $\mathscr{A}$ say, which could be declared C or NC in individual tests and let Z_2' denote the number of items classified as NC on individual testing, among the remaining (n_0-1)

items, so that

$$\Pr[Z'_2 = z \mid Y] = P(z \mid n_0 - 1, Y - 1, p, p'). \tag{8.10}$$

Notice that the NC item $\mathscr{A}$ is classified as NC if (i) $Z_1 > 0$ and either: (ii) $Z'_2 + 1 \geqslant Z_1$, and identification of Z_1 NC items comes before identification of $n_0 - Z_1$ C items, and $\mathscr{A}$ is among the Z_1 NC (with probability $p \min(Z_1/(Z'_2 + 1), 1)$); or (iii) $Z'_2 < Z_1$, and identification of $n_0 - Z_1$ C comes first, and it is among those left without inspection and classified as NC although it would be declared C by individual testing (with probability $(1-p) \times \max((Z'_1 - Z'_2)/(n_0 - Z'_2), 0)$). Therefore, conditional on Z'_2 and Z_1, the probability of correct identification is

$$\mathrm{PC}(\mathrm{NC} \mid Z_1, Z'_2) = p \min\left(\frac{Z_1}{Z'_2 + 1}, 1\right) + (1-p) \max\left(\frac{Z_1 - Z'_2}{n_0 - Z'_2}, 0\right), \tag{8.11}$$

where (i) is satisfied since $Z_1 = 0$ implies PC(NC) = 0. The unconditional probability PC(NC) is obtained by taking the expected value with respect to Z_1 and Z'_2.

$$\mathrm{PC}(\mathrm{NC}) = E^*_Y\{E_{Z_1, Z'_2}[\mathrm{PC}(\mathrm{NC}) \mid Z_1, Z'_2) \mid Y]\}, \tag{8.12}$$

where expectation E^*_Y is with respect to Y, conditioned on $Y \geqslant 1$. Explicitly

$$\mathrm{PC}(\mathrm{NC}) = \sum_y P^*_y \sum_{z_1=1}^{n_0} \sum_{z'_2=0}^{n_0-1} P(z_1 \mid n_0, y, p_1, p'_1) P(z'_2 \mid n_0 - 1, y - 1, p, p')$$
$$\times \left\{ p \min\left(\frac{z_1}{z'_2 + 1}, 1\right) + (1-p) \max\left(\frac{z_1 - z'_2}{n_0 - z'_2}, 0\right) \right\}, \tag{8.13}$$

where $P^*_y = \Pr[Y = y]/(1 - \Pr[Y = 0])$. For a finite lot size

$$P^*_y = \frac{\binom{D-1}{y-1}\binom{N-D}{n_0-y}}{\binom{N-1}{n_0-1}}, \qquad \text{if } \max(1, D + n_0 - N) \leqslant y \leqslant \min(n_0, D) \tag{8.14a}$$

and for an effectively infinite lot size

$$P^*_y = \binom{n_0 - 1}{y - 1} \omega^{y-1} (1 - \omega)^{n_0 - y}, \qquad \text{if } 1 \leqslant y \leqslant n_0. \tag{8.14b}$$

Values of PC(NC), calculated from (8.13), for selected sets of parameter values, are given in Tables 9.2 and 10.2 of Johnson *et al.* (1988) and partially reproduced in section 8.3.

Evaluation of PC(C)

First, we specify a C item, $\mathscr{A}$ say, which could be declared C or NC in individual tests. With the same definition of Z'_2 as in the previous paragraph we now have

$$\Pr[Z'_2 = z \mid Y] = P(z \mid n_0 - 1, Y, p, p'). \tag{8.15}$$

The C item $\mathscr{A}$ is classified correctly if (i) $Z_1 = 0$, or (ii) $Z_1 \geqslant Z'_2$, and identification of $n_0 - Z_1$ C items comes before identification of Z_1 NC items, and $\mathscr{A}$ is among the $n_0 - Z_1$ C items (with probability $(1-p')\min[(n_0 - Z_1)/(n_0 - Z'_2), 1)]$; or (iii) $Z'_2 + 1 \geqslant Z_1$, and identification of Z_1 NC items comes first, and $\mathscr{A}$ is among those left without inspection and classified as C although it would be declared NC in individual tests (with probability $p'\max((Z'_2 - Z_1 + 1)/(Z'_2 + 1), 0))$. Conditioned on Z'_2 and Z_1, the probability of correct identification is therefore

$$\mathrm{PC}(\mathrm{C} \mid Z_1, Z'_2) = (1-p')\min\left(\frac{n_0 - Z_1}{n_0 - Z'_2}, 1\right) + p'\max\left(\frac{Z'_2 - Z_1 + 1}{Z'_2 + 1}, 0\right), \tag{8.16}$$

where, as in the case of the formulas for PC(NC), (i) is satisfied since $Z_1 = 0$ implies $\mathrm{PC(C)} = 1$. The unconditional probability PC(C) is thus:

$$\mathrm{PC(C)} = E^{**}_Y\{E_{Z_1, Z'_2}[\mathrm{PC}(\mathrm{C} \mid Z_1, Z'_2) \mid Y]\}, \tag{8.17}$$

where the expectation E^{**}_Y is with respect to Y, conditioned on $Y \neq n_0$, or

$$\mathrm{PC(C)} = \sum_y P^{**}_y \sum_{z_1=1}^{n_0} \sum_{z'_2=0}^{n_0-1} P(z_1 \mid n_0, y, p_1, p'_1) P(z'_2 \mid n_0 - 1, y, p, p') \times \left\{(1-p')\min\left(\frac{n_0 - z_1}{n_0 - z'_2}, 1\right) + p'\max\left(\frac{1 + z'_2 - z_1}{z'_2 + 1}, 0\right)\right\}, \tag{8.18}$$

where $P_y^{**} = \Pr[Y=y]/(1-\Pr[Y=n_0])$, and for a finite lot size

$$P_y^{**} = \frac{\binom{D}{y}\binom{N-D-1}{n_0-y-1}}{\binom{N-1}{n_0-1}},$$

$$\text{if } \max(0, D+n_0-N) \leqslant y \leqslant \min(n_0-1, D), \quad (8.19a)$$

and, for an effectively infinite lot size,

$$P_y^{**} = \binom{n_0-1}{y}\omega^y(1-\omega)^{n_0-1-y} \qquad \text{if } 0 \leqslant y \leqslant n_0-1. \quad (8.19b)$$

(cf. equations (8.12)–(8.14b)). Values of PC(C), calculated from (8.18), for several sets of parameter values are presented in Tables 9.3 and 10.3 of Johnson *et al.* (1988) and are partially reproduced in Section 8.3.

The overall probability of correct classification, then, is

$$\text{PC} \equiv \begin{cases} \omega\,\text{PC(NC)} + (1-\omega)\text{PC(C)} & \text{for infinite lots;} \\ \{D\cdot\text{PC(NC)} + (N-D)\cdot\text{PC(C)}\}/N & \text{for finite lots.} \end{cases} \quad (8.20)$$

A note on the evaluation of $E[M]$ in the error-free case

If inspection is perfect then $Z_1 = Z_2 = Y$ and from (8.7):

$$E[M|Y] = n_0 + 2 - (n_0+1)\{(Y+1)^{-1} + (n_0-Y+1)^{-1}\}. \quad (8.21)$$

Taking expectations of both sides of (8.21), we have

$$E[M] = n_0 + 2 - (n_0+1)\{E[(Y+1)^{-1}] + E[(n_0-Y+1)^{-1}]\}. \quad (8.22)$$

To obtain $E[M]$ we only need to evaluate $E[(Y+1)^{-1}]$ and $E[(n_0-Y+1)^{-1}]$.

For an infinite lot size (when the distribution of Y is binomial)

$$E[(Y+1)^{-1}] = \frac{1}{(n_0+1)\omega}\left\{1-(1-\omega)^{n_0+1}\right\}; \quad (8.23a)$$

$$E[(n_0-Y+1)^{-1}] = \frac{1}{(n_0+1)(1-\omega)}\left(1-\omega^{n_0+1}\right); \quad (8.23b)$$

therefore

$$E[M]=n_0+2-\{\omega(1-\omega)\}^{-1}\left\{1-\omega^{n_0+2}-(1-\omega)^{n_0+2}\right\}. \quad (8.24)$$

For a finite lot size, (when the distribution of Y is hypergeometric):

$$E[(Y+1)^{-1}] = \begin{cases} \dfrac{N+1}{(n_0+1)(D+1)}\left\{1-\dfrac{(N-n_0)!(N-D)!}{(N+1)!(N-n_0-D-1)!}\right\}, & \text{if } N>n_0+D \\ \dfrac{N+1}{(n_0+1)(D+1)} & \text{if } n_0+D\geqslant N \end{cases} \quad (8.25a)$$

(Johnson and Kotz (1969, p. 145)), and

$$E[(n_0-Y+1)^{-1}] = \begin{cases} \dfrac{N+1}{(n_0+1)(N-D+1)}\left\{1-\dfrac{(N-n_0)!D!}{(N+1)!(D-n_0-1)!}\right\}, & \text{if } D>n_0 \\ \dfrac{N+1}{(n_0+1)(N-D+1)} & \text{if } n_0\geqslant D. \end{cases} \quad (8.25b)$$

Using the second formula in each of (8.25a) and (8.25b) as an approximation,

$$E[M] \doteq n_0+2-(N+1)\left(\frac{1}{N-D+1}+\frac{1}{D+1}\right)$$

$$\doteq n_0+2-\frac{(N+1)(N+2)}{(N-D+1)(D+1)}. \quad (8.26)$$

This approximation is applicable if the sample size (n_0) is large compared with the number of NC items (D) in the population. Additional technical details are given in Kotz, Mao and Johnson (1986).

8.3 Tables and comments

The following tables and figures are based on extensive tables in Johnson *et al.* (1988). Tables 8.1–8.3 give values of $(1+E[M])$,

Table 8.1 *Expected number of tests* $(1 + E[M])$.

$\omega = 0{\cdot}05$, $n = 10$, $p_1 = 0{\cdot}95$, $p'_1 = 0{\cdot}05$

p	$p' = 0$	$p' = 0{\cdot}025$	$p' = 0{\cdot}05$	$p' = 0{\cdot}075$	$p' = 0{\cdot}1$
0·75	5·9152	5·6361	5·3752	5·1314	4·9039
0·80	5·8804	5·6019	5·3752	5·0990	4·8725
0·85	5·8450	5·5672	5·3079	5·0661	4·8406
0·90	5·8089	5·5319	5·2736	5·0328	4·8084
0·95	5·7721	5·4960	5·2387	4·9990	4·7759

$\omega = 0{\cdot}05$, $n = 10$, $p_1 = 0{\cdot}95$, $p'_1 = 0{\cdot}00$

p	$p' = 0$	$p' = 0{\cdot}025$	$p' = 0{\cdot}05$	$p' = 0{\cdot}075$	$p' = 0{\cdot}1$
0·75	3·5352	3·3707	3·2200	3·0819	2·9552
0·80	3·4718	3·3126	3·1668	3·0331	2·9105
0·85	3·4074	3·2538	3·1129	2·9838	2·8654
0·90	3·3421	3·1941	3·0584	2·9340	2·8199
0·95	3·2757	3·1336	3·0032	2·8836	2·7739

$\omega = 0{\cdot}10$, $n = 10$, $p_1 = 0{\cdot}95$, $p'_1 = 0{\cdot}05$

p	$p' = 0$	$p' = 0{\cdot}025$	$p' = 0{\cdot}05$	$p' = 0{\cdot}075$	$p' = 0{\cdot}1$
0·75	6·8925	6·6245	6·3701	6·1287	5·8999
0·80	6·8447	6·5756	6·3205	6·0788	5·8501
0·85	6·7947	6·5248	6·2692	6·0275	5·7989
0·90	6·7425	6·4720	5·9745	5·9745	5·7463
0·95	6·6881	6·4172	5·9200	5·9200	5·6924

$\omega = 0{\cdot}10$, $n = 10$, $p_1 = 0{\cdot}95$, $p'_1 = 0{\cdot}00$

p	$p' = 0$	$p' = 0{\cdot}025$	$p' = 0{\cdot}05$	$p' = 0{\cdot}075$	$p' = 0{\cdot}1$
0·75	5·3212	5·0827	4·8615	4·6563	4·4659
0·80	5·2296	4·9960	4·7795	4·5789	4·3928
0·85	5·1348	4·9066	4·6953	4·4995	4·3181
0·90	5·0368	4·8145	4·6087	4·4182	4·2418
0·95	4·9355	4·7195	4·5198	4·3350	4·1639

Table 8.2 *Values of* PC(NC).

$\omega = 0{\cdot}05,\ n = 10,\ p_1 = 0{\cdot}95,\ p'_1 = 0{\cdot}05$

p	$p' = 0$	$p' = 0{\cdot}025$	$p' = 0{\cdot}05$	$p' = 0{\cdot}075$	$p' = 0{\cdot}1$
0·75	0·7663	0·7213	0·6789	0·6391	0·6018
0·80	0·8063	0·7591	0·7146	0·6728	0·6337
0·85	0·8465	0·7969	0·7503	0·7065	0·6655
0·90	0·8867	0·8348	0·7860	0·7402	0·6973
0·95	0·9271	0·8727	0·8216	0·7737	0·7289

$\omega = 0{\cdot}05,\ n = 10,\ p_1 = 0{\cdot}95,\ p'_1 = 0{\cdot}00$

p	$p' = 0$	$p' = 0{\cdot}025$	$p' = 0{\cdot}05$	$p' = 0{\cdot}075$	$p' = 0{\cdot}1$
0·75	0·7433	0·6790	0·6217	0·5708	0·5254
0·80	0·7844	0·7164	0·6559	0·6020	0·5541
0·85	0·8257	0·7538	0·6900	0·6332	0·5827
0·90	0·8670	0·7912	0·7240	0·6642	0·6110
0·95	0·9085	0·8287	0·7579	0·6950	0·6392

$\omega = 0{\cdot}10,\ n = 10,\ p_1 = 0{\cdot}95,\ p'_1 = 0{\cdot}05$

p	$p' = 0$	$p' = 0{\cdot}025$	$p' = 0{\cdot}05$	$p' = 0{\cdot}075$	$p' = 0{\cdot}1$
0·75	0·7713	0·7343	0·6988	0·6650	0·6328
0·80	0·8099	0·7708	0·7334	0·6978	0·6639
0·85	0·8488	0·8074	0·7680	0·7305	0·6949
0·90	0·8879	0·8441	0·8025	0·7631	0·7257
0·95	0·9272	0·8810	0·8371	0·7956	0·7563

$\omega = 0{\cdot}10,\ n = 10,\ p_1 = 0{\cdot}95,\ p'_1 = 0{\cdot}00$

p	$p' = 0$	$p' = 0{\cdot}025$	$p' = 0{\cdot}05$	$p' = 0{\cdot}075$	$p' = 0{\cdot}1$
0·75	0·7501	0·6982	0·6511	0·6083	0·5693
0·80	0·7898	0·7345	0·6844	0·6389	0·5977
0·85	0·8296	0·7707	0·7175	0·6694	0·6258
0·90	0·8695	0·8069	0·7505	0·6996	0·6536
0·95	0·9097	0·8432	0·7834	0·7295	0·6810

Table 8.3 PC(C).

$\omega = 0{\cdot}05$, $n = 10$, $p_1 = 0{\cdot}95$, $p'_1 = 0{\cdot}05$

p	$p' = 0$	$p' = 0{\cdot}025$	$p' = 0{\cdot}05$	$p' = 0{\cdot}075$	$p' = 0{\cdot}1$
0·75	0·5535	0·5511	0·5489	0·5468	0·5448
0·80	0·5556	0·5531	0·5507	0·5485	0·5465
0·85	0·5577	0.5551	0·5526	0·5503	0·5482
0·90	0·5598	0·5571	0·5545	0·5521	0·5498
0·95	0·5619	0·5591	0·5564	0·5539	0·5515

$\omega = 0{\cdot}05$, $n = 10$, $p_1 = 0{\cdot}95$, $p'_1 = 0{\cdot}00$

p	$p' = 0$	$p' = 0{\cdot}025$	$p' = 0{\cdot}05$	$p' = 0{\cdot}075$	$p' = 0{\cdot}1$
0·75	0·3438	0·3404	0·3374	0·3347	0·3323
0·80	0·3460	0·3424	0·3392	0·3364	0·3338
0·85	0·3481	0·3443	0·3410	0·3380	0·3353
0·90	0·3503	0·3463	0·3428	0·3396	0·3368
0·95	0·3525	0·3483	0·3446	0·3412	0·3383

$\omega = 0{\cdot}10$, $n = 10$, $p_1 = 0{\cdot}95$, $p'_1 = 0{\cdot}05$

p	$p' = 0$	$p' = 0{\cdot}025$	$p' = 0{\cdot}05$	$p' = 0{\cdot}075$	$p' = 0{\cdot}1$
0·75	0·6857	0·6816	0·6776	0·6739	0·6703
0·80	0·6900	0·6856	0·6815	0·6775	0·6738
0·85	0·6943	0·6897	0·6853	0·6811	0·6772
0·90	0·6986	0·6938	0·6892	0·6848	0·6806
0·95	0·7030	0·6979	0·6930	0·6884	0·6840

$\omega = 0{\cdot}10$, $n = 10$, $p_1 = 0{\cdot}95$, $p'_1 = 0{\cdot}00$

p	$p' = 0$	$p' = 0{\cdot}025$	$p' = 0{\cdot}05$	$p' = 0{\cdot}075$	$p' = 0{\cdot}1$
0·75	0·5706	0·5648	0·5596	0·5548	0·5505
0·80	0·5750	0·5688	0·5633	0·5582	0·5536
0·85	0·5794	0·5728	0·5669	0·5616	0·5568
0·90	0·5838	0·5769	0·5706	0·5649	0·5598
0·95	0·5883	0·5809	0·5743	0·5683	0·5629

PC(NC) and PC(C), respectively, for parameter values:

$$N = 100; n_0 = 10; D = 5, 10$$

$$p_1 = 0{\cdot}95; p = 0{\cdot}75(0{\cdot}05)0{\cdot}95; p'_1 = 0, 0{\cdot}05; p' = 0(0{\cdot}025)0{\cdot}1.$$

Tables 8.4–8.6 give values of the same quantities for effectively infinite lots with $\omega = 0{\cdot}05, 0{\cdot}10$ and the same sets of parameter values. From these tables the following observations can be made.

1. The procedure is quite effective in reducing the expected number of tests.
2. The values of $E[M]$ are greatly influenced by changes of $(1-p_1)$ and p'_1, but variations of $(1-p)$ and p' have little influence on $E[M]$ (*no* influence if the procedure is not curtailed) because the expected number mainly depends on results of group testing.
3. Values of PC(NC) and PC(C) vary considerably with variation in p and p'. The effect of p'_1 is also noticeable. The amount of variation decreases as the chance of individual testing decreases, and so decreases with D. Unfortunately, errors in inspection reduce considerably the probabilities of correct decisions, as is to be expected.
4. From $N = 100$ to $N = \infty$, there are only slight changes in all three quantities. Therefore, for lot size $N > 100$ with $D/N = \omega$, these quantities can be interpolated from the tabulated values for $N = 100$ and $N = \infty$. Moreover, it appears, by examining the values for

$$N = 50, \qquad D = 5;$$
$$N = 100, \qquad D = 10;$$
$$N = \infty, \qquad \omega = 0{\cdot}10,$$

that harmonic interpolation (linear with respect to $100/N$) will give good results for $N \geqslant 50$, and probably for N as small as 25 (i.e. $100/N = 4$). For instance, taking parameters $n = 10$, $p_1 = 0{\cdot}80$, $p'_1 = 0{\cdot}05$, $p = 0{\cdot}85$ and $p' = 0{\cdot}10$ we have (Johnson *et al.* 1988) the following tabulated values:

Cases	$1 + E[M]$	PC(NC)	PC(C)	$(100/N)$
$N = 50,\ D = 5$	5·4478	0·5950	0·6813	(2)
$N = 100, D = 10$	5·4162	0·5978	0·6728	(1)
$N = \infty,\ \omega = 0{\cdot}10$	5·3864	0·5991	0·6649	(0)

Table 8.4 *Expected number of tests* $1 + E[M]$.

$N = 100$, $D = 5$, $n = 10$, $p_1 = 0{\cdot}95$, $p'_1 = 0{\cdot}05$

p	$p' = 0$	$p' = 0{\cdot}025$	$p' = 0{\cdot}05$	$p' = 0{\cdot}075$	$p' = 0{\cdot}1$
0·75	5·9571	5·6745	5·4103	5·1635	4·9331
0·80	5·9202	5·6384	5·3753	5·1296	4·9004
0·85	5·8827	5·6019	5·3398	5·0953	4·8673
0·90	5·8446	5·5648	5·3039	5·0606	4·8340
0·95	5·8059	5·5273	5·2676	5·0256	4·8003

$N = 100$, $D = 5$, $n = 10$, $p_1 = 0{\cdot}95$, $p'_1 = 0{\cdot}00$

p	$p' = 0$	$p' = 0{\cdot}025$	$p' = 0{\cdot}05$	$p' = 0{\cdot}075$	$p' = 0{\cdot}1$
0·75	3·6090	3·4356	3·2769	3·1317	2·9988
0·80	3·5421	3·3746	3·2213	3·0810	2·9525
0·85	3·4744	3·3129	3·1651	3·0298	2·9059
0·90	3·4058	3·2505	3·1084	2·9782	2·8589
0·95	3·3364	3·1875	3·0511	2·9261	2·8116

$N = 100$, $D = 10$, $n = 10$, $p_1 = 0{\cdot}95$, $p'_1 = 0{\cdot}05$

p	$p' = 0$	$p' = 0{\cdot}025$	$p' = 0{\cdot}05$	$p' = 0{\cdot}075$	$p' = 0{\cdot}1$
0·75	6·9506	6·6790	6·4209	6·1761	5·9438
0·80	6·8999	6·6274	6·3689	6·1239	5·8919
0·85	6·8472	6·5739	6·3152	6·0703	5·8387
0·90	6·7923	6·5186	6·2598	6·0152	5·7843
0·95	6·7352	6·4613	6·2027	5·9587	5·7283

$N = 100$, $D = 10$, $n = 10$, $p_1 = 0{\cdot}95$, $p'_1 = 0{\cdot}00$

p	$p' = 0$	$p' = 0{\cdot}025$	$p' = 0{\cdot}05$	$p' = 0{\cdot}075$	$p' = 0{\cdot}1$
0·75	5·4228	5·1738	4·9430	4·7291	4·5309
0·80	5·3267	5·0830	4·8575	4·6486	4·4551
0·85	5·2275	4·9897	4·7698	4·5663	4·3779
0·90	5·1251	4·8938	4·6800	4·4822	4·2992
0·95	5·0196	4·7952	4·5879	4·3962	4·2190

Table 8.5 *Values of* PC(NC).

$N = 100$, $D = 5$, $n = 10$, $p_1 = 0{\cdot}95$, $p'_1 = 0{\cdot}05$

p	$p' = 0$	$p' = 0{\cdot}025$	$p' = 0{\cdot}05$	$p' = 0{\cdot}075$	$p' = 0{\cdot}1$
0·75	0·7654	0·7188	0·6750	0·6341	0·5957
0·80	0·8056	0·7568	0·7109	0·6680	0·6278
0·85	0·8460	0·7949	0·7468	0·7018	0·6598
0·90	0·8865	0·8329	0·7827	0·7356	0·6917
0·95	0·9271	0·8711	0·8186	0·7694	0·7235

$N = 100$, $D = 5$, $n = 10$, $p_1 = 0{\cdot}95$, $p'_1 = 0{\cdot}00$

p	$p' = 0$	$p' = 0{\cdot}025$	$p' = 0{\cdot}05$	$p' = 0{\cdot}075$	$p' = 0{\cdot}1$
0·75	0·7420	0·6752	0·6159	0·5634	0·5167
0·80	0·7834	0·7128	0·6503	0·5948	0·5455
0·85	0·8250	0·7505	0·6845	0·6261	0·5742
0·90	0·8666	0·7882	0·7187	0·6572	0·6027
0·95	0·9082	0·8258	0·7529	0·6883	0·6310

$N = 100$, $D = 10$, $n = 10$, $p_1 = 0{\cdot}95$, $p'_1 = 0{\cdot}05$

p	$p' = 0$	$p' = 0{\cdot}025$	$p' = 0{\cdot}05$	$p' = 0{\cdot}075$	$p' = 0{\cdot}1$
0·75	0·7704	0·7323	0·6959	0·6612	0·6282
0·80	0·8093	0·7690	0·7306	0·6941	0·6595
0·85	0·8483	0·8059	0·7654	0·7270	0·6906
0·90	0·8876	0·8428	0·8002	0·7598	0·7216
0·95	0·9271	0·8798	0·8349	0·7925	0·7524

$N = 100$, $D = 10$, $n = 10$, $p_1 = 0{\cdot}95$, $p'_1 = 0{\cdot}00$

p	$p' = 0$	$p' = 0{\cdot}025$	$p' = 0{\cdot}05$	$p' = 0{\cdot}075$	$p' = 0{\cdot}1$
0·75	0·7489	0·6954	0·6469	0·6030	0·5631
0·80	0·7888	0·7318	0·6803	0·6337	0·5915
0·85	0·8289	0·7683	0·7136	0·6643	0·6197
0·90	0·8691	0·8047	0·7468	0·6946	0·6475
0·95	0·9095	0·8412	0·7798	0·7247	0·6751

Table 8.6 *Values of* PC(C).

$N = 100$, $D = 5$, $n = 10$, $p_1 = 0{\cdot}95$, $p_1' = 0{\cdot}05$

p	$p' = 0$	$p' = 0{\cdot}025$	$p' = 0{\cdot}05$	$p' = 0{\cdot}075$	$p' = 0{\cdot}1$
0·75	0·5621	0·5597	0·5574	0·5552	0·5532
0·80	0·5643	0·5617	0·5593	0·5570	0·5549
0·85	0·5664	0·5637	0·5612	0·5588	0·5566
0·90	0·5685	0·5657	0·5631	0·5606	0·5583
0·95	0·5707	0·5677	0·5649	0·5624	0·5599

$N = 100$, $D = 5$, $n = 10$, $p_1 = 0{\cdot}95$, $p_1' = 0{\cdot}00$

p	$p' = 0$	$p' = 0{\cdot}025$	$p' = 0{\cdot}05$	$p' = 0{\cdot}075$	$p' = 0{\cdot}1$
0·75	0·3583	0·3548	0·3517	0·3490	0·3465
0·80	0·3605	0·3568	0·3535	0·3506	0·3480
0·85	0·3627	0·3588	0·3553	0·3522	0·3495
0·90	0·3649	0·3608	0·3571	0·3539	0·3510
0·95	0·3671	0·3628	0·3589	0·3555	0·3525

$N = 100$, $D = 10$, $n = 10$, $p_1 = 0{\cdot}95$, $p_1' = 0{\cdot}05$

p	$p' = 0$	$p' = 0{\cdot}025$	$p' = 0{\cdot}05$	$p' = 0{\cdot}075$	$p' = 0{\cdot}1$
0·75	0·6970	0·6928	0·6888	0·6849	0·6812
0·80	0·7014	0·6969	0·6926	0·6886	0·6847
0·85	0·7057	0·7010	0·6965	0·6922	0·6882
0·90	0·7101	0·7051	0·7003	0·6959	0·6916
0·95	0·7144	0·7092	0·7042	0·6995	0·6950

$N = 100$, $D = 10$, $n = 10$, $p_1 = 0{\cdot}95$, $p_1' = 0{\cdot}00$

p	$p' = 0$	$p' = 0{\cdot}025$	$p' = 0{\cdot}05$	$p' = 0{\cdot}075$	$p' = 0{\cdot}1$
0·75	0·5896	0·5837	0·5783	0·5734	0·5690
0·80	0·5940	0·5877	0·5820	0·5768	0·5721
0·85	0·5985	0·5918	0·5857	0·5802	0·5752
0·90	0·6030	0·5958	0·5894	0·5836	0·5783
0·95	0·6074	0·5999	0·5930	0·5869	0·5814

In each case the value for $N = 100$, $D = 10$ differs from the arithmetic mean of the values for $N = 50$, $D = 5$ and $N = \infty$, $\omega = 0{\cdot}10$ by no more than 0·001.

In regard to 3 above, it is possible that better properties might be obtained if an improved estimator of Y could be used, in place of Z_1. The expected value of Z_1 (conditioned on Y) is

$$Y_{p_1} + (n_0 - Y)p'_1.$$

The conditional expected value of $(Z_1 - n_0 p'_1)/(p_1 - p'_1)$ is Y, but this quantity is not (in general) an integer. One might use

$$\hat{Y} = \text{nearest integer to } (Z_1 - n_0 p'_1)/(p_1 - p'_1),$$

subject to $0 \leqslant \hat{Y} \leqslant n_0$, as a modified estimator of Y, replacing Z_1 in the stopping rules for individual testing. Of course, calculation of $\hat{Y}$ is feasible, only if p_1 and p'_1 are known.

CHAPTER 9

Graff and Roeloffs' approach and related procedures

9.1 Introduction

In previous chapters, there has been emphasis on reduction of the number of tests, and this may increase the error probabilities. For reducing probabilities of misclassification, Graff and Roeloffs (1972, 1974) proposed yet another modification of the Dorfman procedure. The basic idea is to inspect the same item or group repeatedly until we feel that a reasonable decision can be made. Of course this modified procedure can be used only if repetition of inspection is feasible. The procedure is simply repeating inspection of each item or group until either (a) reaching r_1 NC decisions, with final decision NC, or (b) reaching r_2 C decisions with final decision C, whichever ((a) or (b)), comes first. (This is illustrated in Fig. 9.1.) As a consequence of repeating inspection, the two probabilities $(1-p)$, of misclassifying an NC item (or group), and p', of misclassifying a C item (or group), will be reduced to new values $(1-p^*)$ and p'^*, say, respectively. Naturally, this procedure will inevitably increase the expected number of inspections. It is clear that p^* is the probability of obtaining r_1 or more successes, or equivalently obtaining r_2-1 or fewer failures, in r_1+r_2-1 Bernoulli trials with p^* as the success probability. A similar argument applies to p'^*. We have

$$p^* \equiv B(r_1+r_2-1, r_2-1, 1-p) = 1-B(r_1+r_2-1, r_1-1, p); \qquad (9.1\text{a})$$

$$p'^* \equiv B(r_1+r_2-1, r_2-1, 1-p') = 1-B(r_1+r_2-1, r_1-1, p'), \qquad (9.1\text{b})$$

where

$$B(a,b,\theta) = B(a, a-b, 1-\theta) = \sum_{g=0}^{b} \binom{a}{g} \theta^g (1-\theta)^{a-g}. \qquad (0 \leqslant b \leqslant a)$$

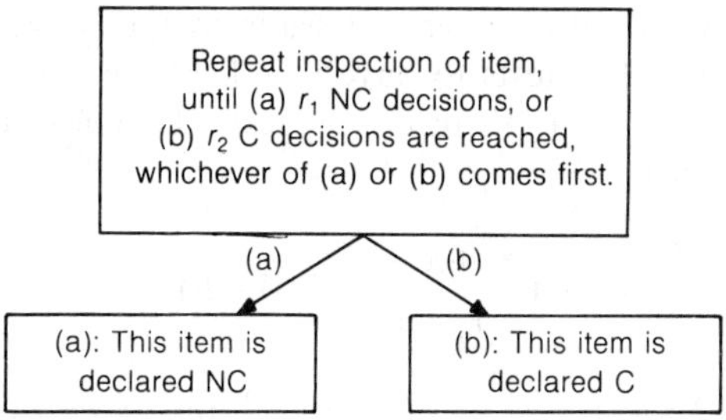

Figure 9.1 *Flow chart for Graff–Roeloffs' procedure*

is the cumulative distribution function of the binomial distribution with parameters (a, θ).

9.2 Graff and Roeloffs' modification

In order to obtain the formulas for the expected number of inspections, we first consider a rather general argument. Let A be an event, and A^c be its complement. We now consider a sequence of tests in which we will stop testing as soon as r_1 A's or r_2 A^c's have been observed. Let $\theta = \Pr[A]$. The expected number of tests is

$$\sum_{x=0}^{r_2-1} (r_1+x)\binom{r_1+x-1}{r_1-1}\theta^{r_1}(1-\theta)^x + \sum_{x=0}^{r_1-1} (r_2+x)\binom{r_2+x-1}{r_2-1}(1-\theta)^{r_2}\theta^x, \tag{9.2a}$$

where the two terms correspond to stopping with $r_1 A$'s and $r_2 A^c$'s, respectively. The first term, in fact, equals

$$\theta^{-1}r_1 \sum_{x=0}^{r_2-1} (r_1+x)\binom{r_1+x}{r_1}\theta^{r_1+1}(1-\theta)^x.$$

Note also that

$$\binom{r_1+x}{r_1}\theta^{r_1+1}(1-\theta)^x$$
$$= \Pr[\text{need exactly } r_1+x+1 \text{ tests to get } r_1+1 A\text{'s}].$$

Summing over $x=0$ to $x=r_2-1$ leads to the probability that it takes at most r_1+r_2 tests to have r_1+1 A's; in other words, that there are at lest r_1+1 A's in r_1+r_2 tests. Therefore the first term of (9.2a) becomes

$$\begin{aligned}
&\sum_{x=0}^{r_2-1}(r_1+x)\binom{r_1+x-1}{r_1-1}\theta^{r_1}(1-\theta)^x \\
&=\theta^{-1}r_1\sum_{x=0}^{r_2-1}(r_1+x)\binom{r_1+x}{r_1}\theta^{r_1+1}(1-\theta)^x \\
&=\theta^{-1}r_1\{1-B(r_1+r_2,r_1,\theta)\} \\
&=\theta^{-1}r_1B(r_1+r_2,r_2-1,1-\theta).
\end{aligned}$$

The last equality is a direct consequence of properties of the binomial distribution. Applying a similar argument to the second term of (9.2a), we obtain the expected number of tests

$$r_2(1-\theta)^{-1}B(r_1+r_2,r_1-1,\theta)+r_1\theta^{-1}B(r_1+r_2,r_2-1,1-\theta). \quad (9.2b)$$

If A now denotes the event 'declaring NC correctly,' then $\theta=p$; if A^c denotes the event 'declaring C correctly,' then $\theta=p'$, Replacing θ by p and p' in (9.2a) and (9.2b) leads to (9.3a) and (9.3b) below, respectively.

1. If at least one NC item is present:

$$E=\begin{cases} r_2(1-p)^{-1}B(r_1+r_2,r_1-1,p) \\ \quad +r_1p^{-1}B(r_1+r_2,r_2-1,1-p) & \text{for } p<1; \\ r_1 & \text{if } p=1; \end{cases} \quad (9.3a)$$

2. If no NC items are present:

$$E'=\begin{cases} r_2(1-p')^{-1}B(r_1+r_2,r_1-1,p') \\ \quad +r_1p'^{-1}B(r_1+r_2,r_2-1,1-p') & \text{for } p'>0; \\ r_1 & \text{if } p'=0. \end{cases} \quad (9.3b)$$

(Note that in (9.3a), with $p=1$, every test gives a (correct) NC decision, and in (9.3b) with $p'=0$, every test gives a (correct) C decision.)

Modification of the Dorfman procedure

We apply the modification to the original Dorfman procedure. To obtain formulas for PC(NC) and PC(C), the probabilities of correct classification of an NC item and C item, respectively, we just use p^* (p_0^* for group testing) and p'^* ($p_0'^*$) in place of p (p_0) and p' (p'_0), respectively, in related formulas for the Dorfman procedure (cf. section 6.2). Note that n_0 is the group size, taken from a lot of size N containing D NC items. We have:

1. probability of correct classification for an NC item is

$$\text{PC(NC)} = p_0^* p^* \tag{9.4}$$

2. probability of correct classification for a C item is

$$\begin{aligned}\text{PC(C)} &= P_0^*(n_0)(1 - p_0'^* p'^*) + (1 - P_0^*(n_0))(1 - p_0^* p'^*) \\ &= 1 - (p_0^* - P_{(0)}^*)p'^*,\end{aligned} \tag{9.5}$$

where

$$\begin{aligned}P_{(0)}^* &\equiv (p_0^* - p_0'^*)P_0^*(n_0); \\ P_0^*(n_0) &\equiv (N - D - 1)^{(n_0 - 1)}/(N - 1)^{(n_0 - 1)},\end{aligned}$$

which is the probability that all items in the sample of size n_0 are C, given that a specified one of them is C.

9.3 Modification for hierarchal screening

Introduction and notation

In this section, Graff and Roeloffs' modification will be applied to the hierarchal Dorfman procedures introduced in section 6.3. Suppose we want to screen a group of size n_0. If the group test gives an NC result, we split it into h_1 subgroups of size n_1 each ($n_0 = h_1 n_1$). Each subgroup is then tested again, and so on. Generally in stage j, we test a sub^{j}group of size n_j where $n_j = n_{j-1}/h_j = n_0/(\Pi_{i=1}^{j} h_i)$. If the sub^{j}group is identified as C (negative result), then all its items are declared C; if it is declared NC, then we test each of the h_{j+1} sub^{j+1} groups, in the sub^{j}group and so forth. As usual, let the lot under consideration have size N with $D(>0)$ NC items, or have infinite size with proportion ω of NC items. We will consider a $(k + 1)$-stage hierarchal screening procedure with Graff and Roeloffs' modification. As before, we will be mainly interested in the three measures P(NC), PC(C) and E.

We allow for changes in the values of r_1 and r_2 through the different stages; for the jth stage we will use r_{1j} and r_{2j} instead, where $j = 1, \ldots, k+1$. Symbols such as p_j^*, $p_j'^*$, E_j^* and $E_j'^*$ are also used in the same way.

Analysis

We obtain probabilities of correct classification PC(NC) and PC(C) simply by replacing p_j by p_j^* and p_j' by $p_j'^*$ in equations (6.15) and (6.16) in Chapter 6, where, from (9.1a) and (9.1b),

$$p_j^* = B(r_{1j} + r_{2j} - 1, r_{2j} - 1, 1 - p_j)$$

and

$$p_j'^* = B(r_{1j} + r_{2j} - 1, r_{2j} - 1, 1 - p_j').$$

The resulting formulas are

$$\text{PC(NC)} = \prod_{j=0}^{k+1} p_j^* \tag{9.6}$$

and

$$\text{PC(C)} = 1 - \sum_{j=0}^{k+1} \{P_0^*(n_j) - P_0^*(n_{j-1})\}\left(\prod_{i=0}^{j-1} p_i^*\right)\left(\prod_{i=j}^{k+1} p_i'^*\right), \tag{9.7}$$

where

$$P_0^*(n) = \begin{cases} (1-\omega)^{n-1} & \text{for infinite lots,} \\ \dfrac{(N-D-1)^{(n-1)}}{(N-1)^{(n-1)}} & \text{for finite lots;} \end{cases}$$

Note that, by convention,

$$P_0^*(n_{-1}) = 0 \quad \text{and} \quad P_0^*(n_{k+1}) = 1.$$

Formula (9.6) follows from the fact that an NC item is correctly classified if and only if correct NC decisions are obtained at each stage of the procedure in which it is involved. Formula (9.7) is obtained by applying the total probability rule, noting that: (a) given a C item in a set of size n_0, $P_0^*(n_s) - P_0^*(n_{s-1})$ is the probability that a *given* sub^{s-1}group of size n_{s-1} contains at least one NC item but a *particular* sub^{s}group does not have any NC item; and (b) conditioned on this situation, the last two products in (9.7) $(\prod_{j=0}^{s-1} p_j^*)(\prod_{j=2}^{k+1} p_j'^*)$ give the probability that this C item is declared NC.

We now derive a formula for the expected number of tests, E. Recall section 6.2, which gives the formula for the expected number of tests in a $(k+1)$-stage hierarchal Dorfman procedure without repetitions:

$$E = \sum_{s=0}^{k+1} \left(\prod_{j=1}^{s} h_j \right) \pi(s), \tag{6.17}$$

where

$$\pi(s) = P_0(n_0) \prod_{i=0}^{s-1} p_i' + \prod_{j=1}^{s-1} \{P_0(n_j) - P_0(n_{j-1})\} \left(\prod_{i=0}^{j-1} p_i \right) \left(\prod_{i=j}^{s-1} p_i' \right)$$
$$+ \{1 - P_0(n_{s-1})\} \prod_{i=0}^{s-1} p_i, \tag{6.18}$$

with $\pi(0) = \prod_{j=1}^{0} h_j = 1$ and $P_0(n)$, the probability that a sample of n items contains no NC items,

$$P_0(n) = \begin{cases} (1-\omega)^n & \text{for infinite lots;} \\ (N-D)^{(n)}/N^{(n)} & \text{for finite lots.} \end{cases}$$

Note that $\prod_{i=0}^{s-1} p_i'$ is the probability that a sub^{s}group is tested if all items in the group (sub^{0}group) are really C, and the product

$$\left(\prod_{i=0}^{j-1} p_i \right) \left(\prod_{i=j}^{s-1} p_i' \right)$$

has a similar meaning to $(\prod_{i=0}^{j-1} p_i^*)(\prod_{i=j}^{k+1} p_i'^*)$ above.

To obtain the expected number of tests needed to reach a decision, NC or C, for a sub^{j}group, we use (9.2a) and (9.2b) with an additional subscript, j, to indicate the particular group, and arrive at

1. if at least one NC item is present:

$$E_j = \begin{cases} r_{2j}(1-p_j)^{-1} B(r_{1j}+r_{2j}, r_{1j}-1, p_j) \\ \quad + r_{1j} p_j^{-1} B(r_{1j}+r_{2j}, r_{2j}-1, 1-p_j), & \text{for } p_j < 1; \\ r_{1j} & \text{for } p_j = 1; \end{cases}$$

Table 9.1 *Properties of Graff–Roeloff's procedures* $n_0 = 12$; $p = 0{\cdot}95$; $p' = 0{\cdot}05$

				$\omega = 0{\cdot}05$						$\omega = 0{\cdot}10$					
				EPR(%)			PC(C)			EPR(%)			PC(C)		
				$k=0$	$k=1$	$k=2$	$k=0$	$k=1$	$k=2$	$k=0$	$k=1$	$k=2$	$k=0$	$k=1$	$k=2$
n_1	n_2	r_{10}	r_{20}	($r_{ij} = 1$ for $j \geqslant 1$)											
3	–	1	1	45·3	61·7	–	0·978	0·995	–	22·1	41·8	–	0·967	0·990	–
		2	1	41·4	60·3	–	0·980	0·995	–	16·3	39·0	–	0·969	0·991	–
		1	2	40·8	54·8	–	0·976	0·994	–	19·6	36·3	–	0·964	0·990	–
		2	2	36·2	52·2	–	0·978	0·995	–	13·0	30·9	–	0·966	0·990	–
4	2	1	1	45·3	61·9	62·2	0·978	0·993	0·998	22·1	41·8	40·8	0·967	0·987	0·995
		2	1	41·4	60·3	60·4	0·980	0·993	0·998	16·3	38·5	37·9	0·969	0·987	0·996
		1	2	40·8	55·2	55·5	0·976	0·992	0·997	19·6	36·0	35·3	0·964	0·986	0·995
		2	2	36·2	52·2	52·4	0·978	0·993	0·995	13·0	30·4	29·7	0·966	0·986	0·995
6	2	1	1	45·3	59·0	62·2	0·978	0·989	0·997	22·1	36·5	40·7	0·967	0·981	0·995
		2	1	41·4	57·3	60·4	0·980	0·990	0·998	16·3	33·8	37·7	0·969	0·982	0·995
		1	2	40·8	52·3	55·9	0·976	0·989	0·997	19·6	31·0	35·4	0·964	0·980	0·995
		2	2	36·2	48·9	52·3	0·978	0·989	0·997	13·0	25·2	29·5	0·966	0·980	0·995
6	3	1	1	45·3	59·0	62·8	0·978	0·989	0·996	22·1	36·5	41·4	0·967	0·981	0·995
		2	1	41·4	57·3	60·8	0·980	0·990	0·996	16·3	33·8	38·3	0·969	0·982	0·992
		1	2	40·8	55·3	56·4	0·976	0·989	0·995	19·6	31·0	36·1	0·964	0·980	0·991
		2	2	36·2	48·9	52·8	0·978	0·989	0·995	13·0	25·2	30·2	0·966	0·980	0·991

n_1	n_2	r_1	r_2	($r_{ij}=r_1$, $r_{2j}=r_2$ (all j))											
3	–	2	1	40·0	53·4	–	0·999	1·000	–	9·7	25·2	–	0·998	1·000	–
		1	2	−8·0	28·0	–	0·953	0·987	–	−46·8	−3·7	–	0·930	0·977	–
		2	2	−13·9	20·4	–	0·997	0·999	–	−67·1	−24·0	–	0·995	0·998	–
4	2	2	1	40·0	53·5	47·2	0·999	1·000	1·000	9·7	25·0	12·2	0·998	0·999	1·000
		1	2	−8·0	28·7	32·5	0·953	0·983	0·994	−46·8	−4·8	−0·6	0·930	0·969	0·988
		2	2	−13·9	19·7	8·1	0·997	0·999	1·000	−67·1	−26·4	−50·3	0·995	0·998	0·999
6	2	2	1	40·0	50·6	47·6	0·999	1·000	1·000	9·7	20·7	13·2	0·998	0·999	1·000
		1	2	−8·0	22·9	32·5	0·953	0·976	0·993	−46·8	−15·2	−2·4	0·930	0·957	0·987
		2	2	−13·9	11·5	7·8	0·997	0·998	1·000	−67·1	−39·5	−51·0	0·995	0·997	0·999
6	3	2	1	40·0	50·6	44·7	0·999	1·000	1·000	9·7	20·7	8·0	0·998	0·999	1·000
		1	2	−8·0	22·9	33·2	0·953	0·976	0·989	−46·8	−15·2	−1·6	0·930	0·957	0·979
		2	2	−13·9	11·5	−2·2	0·997	0·998	0·999	−67·1	−39·5	−69·3	0·995	0·997	0·999

Values of PC(NC) for $p=0{\cdot}95$ (not dependent on p' or ω_j or on n_1 or n_2)

		$k=0$	$k=1$	$k=2$			$k=0$	$k=1$	$k=2$
r_1	r_2	($r_{ij}=r_1, r_{2j}=r_2$ (all j))			r_{10}	r_{20}	($r_{ij}=1$ for $j\geqslant 1$)		
2	1	0·815	0·735	0·664	1	1	0·9025	0·8574	0·8145
1	2	0·995	0·993	0·990	2	1	0·8574	0·8145	0·7738
2	2	0·986	0·978	0·971	1	2	0·9426	0·9003	0·8553
					2	2	0·9430	0·8961	0·8512

2. if no NC item is present:

$$E'_j = \begin{cases} r_{2j}(1-p'_j)^{-1}B(r_{1j}+r_{2j}, r_{1j}-1, p'_j) & \\ \quad + r_{1j}p'^{-1}_j B(r_{1j}+r_{2j}, r_{2j}-1, 1-p'_j), & \text{for } p'_j > 0; \\ r_{2j} & \text{for } p'_j = 0. \end{cases}$$

The overall expected number of inspections is then

$$\begin{aligned} E &= P_0(n_0)E'_0 + (1-P_0(n_0))E_0 \\ &\quad + h_1[\{P_0(n_0)p'^*_0 + (P_0(n_1) - P_0(n_0))p^*_0\}E'_1 + (1-P_0(n_1))p^*_0 E_1] \\ &\quad + h_1 h_2[\{P_0(n_0)p'^*_0 p'^*_1 + (P_0(n_1) - P_0(n_0))p^*_0 p'^*_1 \\ &\quad + (P_0(n_2) - P_0(n_1))p^*_0 p^*_1\}E'_2 + (1-P_0(n_2))p^*_0 p^*_1 E_2] + \cdots \\ &= \sum_{i=1}^{k+1}\left(\prod_{j=1}^{i} h_j\right)\left[\left\{\sum_{u=0}^{i}((P_0(n_u) - P_0(n_{u-1})))\prod_{v=0}^{u-1} p^*_v \prod_{w=u+1}^{i-1} p'^*_w\right\}E'_i \right. \\ &\quad \left. + (1-P_0(n_i))\left(\prod_{v=0}^{i-1} p^*_v\right)E_i\right]. \end{aligned}$$

Although this formula appears quite complicated, it has a simple structure directly summing the probabilities that each sub^{j}group will be tested as a set. Some additional details on derivation are presented in Kotz *et al.* (1988).

Tables

Kotz *et al.* (1988) give tables of values of EPR(%) $= 100(1 - E/n_0)$ (Tables 1, 1A), PC(C) (Tables 2, 2A) and PC(NC) (Tables 3, 3A) for Graff–Roeloffs modified hierarchal Dorfman procedures (two or more stages) with combinations of parameter values:

$$r_{10} = r_{20} = 2; \quad r_{ij} = 1 \text{ (for } j > 0) \quad \text{(Tables 1–3);}$$
$$r_{1j} = r_1 = 1, 2; \quad r_{2j} = r_2 = 1, 2 \quad \text{(Tables 1A–3A);}$$
$$p_j = p = 0{\cdot}90, 0{\cdot}95; \quad p'_j = p' = 0{\cdot}05, 0{\cdot}10;$$

and

$$(p, p') = (0{\cdot}80, 0{\cdot}10), (0{\cdot}80, 0{\cdot}15), (0{\cdot}75, 0{\cdot}20);$$

the population proportion of NC items, $\omega = 0{\cdot}05, 0{\cdot}1$;

group size, $n_0 = 12$.

We reproduce, in Table 9.1, the values for $(p, p') = (0{\cdot}95, 0{\cdot}05)$ only.

(Note that a negative EPR value reflects an increase in expected number of tests.)

Discussion and conclusions

Effects of the modified procedure have to be assessed by balancing the effects of reduced error probabilities against increased expected numbers of tests. Whether it is beneficial or not depends on the cost structure, which includes the cost of testing and the risk of faulty decision. Graff and Roeloffs' modification seems to provide a flexible way to balance values of the three critical quantities PC(C), PC(NC) and E, depending on the purpose of a particular study. An important feature is that it is possible to achieve high values of PC(NC), which is of special concern in many medical applications.

CHAPTER 10

Binary search

10.1 Introduction

As in Chapters 6–9, suppose it is possible by testing a group of items to determine whether there is at least one NC item among them. If a set has 2^k items, and no more than one of them is NC, the following method, based on a very simple idea, can be used to determine which item is NC.

1. First test all 2^k items as a group. If the presence of at least one NC item is indicated, proceed to step 2, otherwise conclude the procedure.
2. Split the set into two subsets, each of size 2^{k-1}. One of the sets, chosen at random, is tested as a group. If the presence of an NC item is indicated, the other set is discarded; otherwise the set tested is discarded because that NC item must be contained in the other subset. Therefore, only one subset is left for further testing.
3. The process continues, starting by dividing the surviving set of 2^{k-1} items into two subsets of 2^{k-2} items each. We could repeat the same kind of search, and after $(k-1)$ stages, just two items would remain.
4. The final, kth, stage consists of testing one of these two items to decide which one is the NC item.

Just k tests are required by this procedure, once step 2 is reached.

If testing is perfect, and there is just one NC item present, it will be isolated with certainty. However, as before, we will assume the inspection is not error-free and study the properties of the binary (successive halving) method, allowing for two types of error. We will still use $1-p$ (resp. p') to denote the probabilities of incorrectly classifying an NC (resp. C) item or group. We also assume that tests are mutually independent, and the values of p and p' do not depend on the size of the set.

More generally, we assume there are really y NC items among the 2^k in the original set. The expected number of tests is $1 + kp$ for $y > 0$ and $1 + kp'$ for $y = 0$. We use $P(y, 2^k)$ to denote the probability that, if the halving procedure starts the item finally selected is indeed NC. Evaluating this probability is our main target. Naturally,

$$P(0, 2^k) = 0 \qquad \text{and} \qquad P(2^k, 2^k) = 1. \tag{10.1}$$

10.2 Evaluation of $P(y, 2^k)$

Calculation of $P(1, 2^k)$

First consider the case $y = 1$, i.e. there is just one NC item. One of the two subsets contains the NC item and the other is clean. Each subset will be tested with probability $\frac{1}{2}$. After the testing the probability of correct classification is p or $1 - p'$, depending on which set is tested, and correct classification means the survival of the NC item to the next stage. The probability that the subset containing the NC item will be retained (or the other one eliminated) to the next stage, which is a measure of the accuracy of the inspection, is

$$\phi = \tfrac{1}{2}p + \tfrac{1}{2}(1 - p') = \tfrac{1}{2}(1 - p' + p). \tag{10.2}$$

If the inspection is perfect, i.e. $p = 1$ and $p' = 0$, then $\phi = 1$; if at the other extreme; it is always wrong, $p = 0$ and $p' = 1$, and $\phi = 0$.

It is obvious that the above analysis is valid for any stage and any k. The probability that some NC item survives all the k stages and is finally selected is

$$P(1, 2^k) = \phi^k, \tag{10.3}$$

General case

The probability that a *specific* NC item will be finally (correctly) selected is

$$\mathrm{PC(NC)} = p\,P(y, 2^k)/y, \tag{10.4}$$

since there is probability p of starting the halving procedure with y NC items present. The value of PC(C), the probability of correct classification of a C item, is obtained in the following way.

1. $y = 0$ *case*. The probability that a given C item is finally selected as 'the NC item' equals the product of two probabilities: one is

the probability that the whole group is (wrongly) indicated NC ($=p'$) and the other is the probability of being the one chosen from all 2^k competitors ($=2^{-k}$). We therefore have $1-\mathrm{PC(C)}=p'2^{-k}$ or

$$\mathrm{PC(C)} = 1 - p'2^{-k}. \tag{10.5a}$$

2. $y > 0$ *case.* Since $1 - P(y, 2^k)$ is the probability that the finally chosen one is not truly NC, p' is the probability of incorrectly classifying C as NC, and $2^k - y$ is the number of C competitors $1 - \mathrm{PC(C)} = p'(2^k - y)^{-1}\{1 - P(y, 2^k)\}$, so that

$$\mathrm{PC(C)} = 1 - p(2^k - y)^{-1}\{1 - P(y, 2^k)\}. \tag{10.5b}$$

Evaluation of $P(y, 2^k)$

Suppose that at any general stage with a surviving set of size 2^k, there are $y(>1)$ surviving NC items. A halving will split these y NC items into two groups, with r and $y-r$ NC items, respectively. This will be called a $(r, y-r)$ **split** with, possibly, $r = 0, \ldots, y$. The probability of a $(r, y-r)$ split, with the subset size 2^h, is

$$p_{r,y-r}(2^h) = \binom{y}{r}\binom{2^h - y}{2^{h-1} - r}\bigg/\binom{2^h}{2^{h-1}}. \tag{10.6}$$

Since there is symmetry between the $(r, y-r)$ and $(y-r, r)$ splits:

$$p_{r,y-r}(2^h) = p_{y-r,r}(2^h). \tag{10.7}$$

Since the probabilities p and p' do not depend on the size of subset, any pair of sets are in equal positions as long as both of them contain at least one NC item. Hence we need consider only two cases: (i) if $r=0$ or $r=y$, i.e. all NC items are in the same set, and the probability that this set survives to the next stage is ϕ; (ii) if $1 \leqslant r \leqslant y-1$, then the two sets are in equal position. Then with probability $\frac{1}{2}$, any particular one of the two sets survives. Considering (10.7), we have

$$P(y, 2^h) = 2p_{y,0}(2^h)\,\phi\, P(y, 2^{h-1}) + \sum_{r=1}^{y-1} p_{r,y-r}(2^h)\, p(r, 2^{h-1}). \tag{10.8}$$

This recursive formula enables us to express $P(y, 2^k)$ in terms of $P(1, 2^k)$'s, which are given by (10.3).

We first note that

$$P(0, 2^k) = 0; \quad P(1, 2^k) = \phi^k; \quad P(2^k, 2^k) = 1.$$

We now give some formulas for evaluating $P(y,2^k)$ for $y=2,\dots,8$, and certain suitable k's for these values of y. As an example, we illustrate the procedure by obtaining $P(y,2^k)$ for $y=2$.

First, from (10.6),

$$p_{0,2}(2^k)=p_{2,0}(2^k)=\tfrac{1}{2}(2^{k-1}-1)(2^k-1)^{-1};$$
$$p_{1,1}(2^k)=2^{k-1}(2^k-1)^{-1},$$

and

$$\begin{aligned}P(2,2^k)&=2p_{2,0}(2^k)\phi P(2,2^{k-1})+p_{1,1}(2^k)P(1,2^{k-1})\\&=2p_{2,0}(2^k)\phi\{2p_{2,0}(2^{k-1})\phi P(2,2^{k-2})\\&\quad+p_{1,1}(2^{k-1})P(1,2^{k-2})\}+p_{1,1}(2^k)P(1,2^{k-1}).\end{aligned}$$

For $y=2$ we obtain

$$p_{2,0}(2^3)=\tfrac{3}{14};\qquad p_{1,1}(2^3)=\tfrac{4}{7};\qquad p_{2,0}(2^2)=\tfrac{1}{6};\qquad p_{1,1}(2^2)=\tfrac{2}{3};$$

and further

$$\begin{aligned}P(2,2^3)&=2\times\tfrac{3}{14}P(2,2^2)+\tfrac{4}{7}P(1,2^2)\\&=2\times\tfrac{3}{14}\phi\{2\times\tfrac{1}{6}\phi P(2,2)+\tfrac{2}{3}P(1,2)\}+\tfrac{4}{7}P(1,2^2)\\&=\tfrac{3}{7}\phi(\tfrac{1}{3}\phi+\tfrac{2}{3}\phi)+\tfrac{4}{7}\phi^2=\phi^2.\end{aligned}$$

Mathematical induction shows that

$$P(2,2^k)=\phi^{k-1},\qquad(k=1,2,\dots).\tag{10.9}$$

Similarly we have

$$P(3,2^k)=\tfrac{1}{2}\phi^{k-2}(1+\phi)\qquad(k=2,3,\dots).\tag{10.10}$$

For $y=4,\dots,8$ and some k, we have the following special formulas:

$$P(4,2^3)=\tfrac{1}{35}(4+23\phi+8\phi^2);$$
$$P(4,2^4)=\tfrac{1}{91}\phi(12+55\phi+24\phi^2);$$
$$P(4,2^5)=\tfrac{1}{29}\phi^2(4+17\phi+8\phi^2);$$
$$P(4,2^6)=\tfrac{1}{427}\phi^3(60+247\phi+120\phi^2);$$
$$P(4,2^7)=\tfrac{1}{875}\phi^4(124+503\phi+248\phi^2);$$
$$P(5,2^3)=\tfrac{1}{14}(4+9\phi+\phi^2);$$
$$P(5,2^4)=\tfrac{1}{546}(8+148\phi+319\phi^2+71\phi^3);$$
$$P(5,2^5)=\tfrac{1}{406}\phi(8+108\phi+229\phi^2+61\phi^3);$$
$$P(5,2^6)=\tfrac{1}{7{,}686}\phi^2(1{,}408+2{,}028\phi+3{,}029\phi^2+1{,}221\phi^3);$$
$$P(5,2^7)=\tfrac{1}{133{,}350}\phi^2(4{,}288+35{,}052\phi+72{,}293\phi^2+21{,}717\phi^3);$$

$P(6,2^3) = \frac{1}{2}(1+\phi)$;

$P(6,2^4) = \frac{1}{2{,}002}(95 + 793\phi + 1{,}002\phi^2 + 112\phi^3)$;

$P(6,2^5) = \frac{1}{133{,}274}(128 + 6{,}063\phi + 41{,}997\phi^2 + 56{,}238\phi^3 + 848\phi^4)$;

$P(6,2^6) = \frac{1}{669{,}414}\phi(1{,}152 + 38{,}137\phi + 241{,}635\phi^2 + 329{,}690\phi^3 + 58{,}800\phi^4)$;

$P(7,2^4) = \frac{1}{2{,}860}(252 + 1{,}417\phi + 1{,}135\phi^2 + 56\phi^3)$;

$P(7,2^5) = \frac{1}{420{,}732}(1{,}720 + 41{,}156\phi + 188{,}413\phi^2 + 173{,}371\phi^3 + 16{,}702\phi^4)$;

$P(8,2^4) = \frac{1}{6{,}435}(948 + 3{,}607\phi + 1{,}848\phi^2 + 32\phi^3)$;

$P(8,2^5) = \frac{1}{2{,}629{,}575}(24{,}392 + 393{,}740\phi + 1{,}305{,}547\phi^2 + 860{,}104\phi^3 + 45{,}792\phi^4)$.

Table 1 of Johnson and Kotz (1988)

Table 10.1, reproduces Table 1 of Johnson and Kotz and gives numerical values of $P(y,2^k)$ for $y = 1,\ldots,8$ and $k = 1,\ldots,6$ with various error probabilities which are represented by different values of ϕ (here $\phi = 0{\cdot}9, 0{\cdot}8$). In assessing the values of $P(y,2^k)$ (and $pP(y,2^k)/y$), it should be kept in mind that the successive halving procedure is not designed for cases where y exceeds 1. The table indicates that $P(y,2^k)$ increases slowly with y for fixed k, but decreases slowly as k increases, with y fixed. Of course, large ϕ corresponds to large $P(y,2^k)$.

An alternative recursion formula

If it is desired to calculate $P(y,2^k)$ when there are only a few conforming items, it is convenient to introduce and evaluate

$$Q(u,2^k) \equiv 1 - P(2^k - u, 2^k). \tag{10.11}$$

When $u = 1$, $Q(1,2^k) = 1 - P(2^k - 1, 2^k)$ is the probability that given just one C item among 2^k items, successive halving will produce that C item. At each of the first $(k-1)$ stages there will be $2^{r-1} - 1$ NC items and one C item in one half and 2^{r-1} NC's in the other half ($r = k, k-1, \ldots, 2$). Both halves contain at least one NC, so each is equally likely to be eliminated. The probability that the part containing the C item is chosen at *every* stage is $(\frac{1}{2})^{k-1}$. If this happens, there will be one C and one NC item remaining. The chance that the NC item is chosen is ϕ, so the chance that the C item is chosen

Table 10.1 *Values of* $P(y, 2^k)$.

y	ϕ	$k=$ / $n=2^k=$	2 / 4	3 / 8	4 / 16	5 / 32	6 / 64
1	0·9		0·810	0·729	0·656	0·590	0·527
	0·8		0·640	0·512	0·410	0·327	0·262
2	0·9		0·900	0·810	0·729	0·656	0·590
	0·8		0·800	0·640	0·512	0·410	0·327
3	0·9		0·950	0·855	0·7695	0·693	0·623
	0·8		0·900	0·720	0·576	0·461	0·369
4	0·9		1·000	0·889	0·801	0·720	0·648
	0·8		1·000	0·789	0·627	0·501	0·401
5	0·9			0·922	0·827	0·743	0·668
	0·8			0·846	0·672	0·536	0·429
6	0·9			0·950	0·850	0·763	0·686
	0·8			0·900	0·713	0·567	0·453
7	0·9			0·975	0·870	0·780	
	0·8			0·950	0·748	0·596	
8	0·9			1·000	0·888	0·796	
	0·8			1·000	0·782	0·622	

NOTE: (1) The probability that an NC item will be selected as a consequence of the entire procedure is $pP(y, 2^k)$.
(2) The probability that a *specified* NC item will be selected is $pP(y, 2^k)/y$.
(3) If $\phi = 1$ (error-free testing), then for $y > 0$, $P(y, 2^k) = pP(y, 2^k) = 1$.

is $(1 - \phi)$. Hence

$$Q(1, 2^k) = (\tfrac{1}{2})^{k-1}(1 - \phi),$$

so

$$P(2^k - 1, 2^k) = 1 - (\tfrac{1}{2})^{k-1}(1 - \phi).$$

The general recursion formulas are

$$Q(u, 2^k) = \sum_{r=1}^{u} P_{r,u-r}(2^k)Q(r, 2^{k-1}) \qquad \text{for } u < 2^{k-1}, \quad (10.12a)$$

and

$$Q(2^{k-1}, 2^k) = \sum_{r=1}^{2^{k-1}} p_{r,2^{k-1}-r}(2^k)Q(r, 2^{k-1}) + p_{0,2^{k-1}}(2^k)(1 - \phi). \qquad (10.12b)$$

Using (10.12a) and mathematical induction we obtain

$$Q(2, 2^k) = (\tfrac{1}{2})^{k-2}(1-\phi),$$

so

$$P(2^k - 2, 2^k) = 1 - (\tfrac{1}{2})^{k-2}(1-\phi), \qquad (k \geqslant 2).$$

For $u > 2^{k-1}$, equation (10.6) is more convenient to use.

10.3 Evaluation of P(y, n), $n \neq 2^k$

The error-free case has been discussed by Sandelius (1961), who derived optimal procedures for these situations.

Formula (10.6) holds for n even, and we only need to replace 2^k by n and 2^{k-1} by $\tfrac{1}{2}n$. For n odd, the split will be $(\tfrac{1}{2}(n-1), \tfrac{1}{2}(n+1))$, and a modification of (10.6) gives:

$$\begin{aligned} P(y,n) = {} & \{p_{y,0}(n)P(y, \tfrac{1}{2}(n-1)) + p_{0,y}(n)P(y, \tfrac{1}{2}(n+1))\}\phi \\ & + \tfrac{1}{2}\sum_{r=1}^{y-r} p_{r,y-1}(n)\{P(r, \tfrac{1}{2}(n-1)) + P(y-r, \tfrac{1}{2}(n+1))\}, \end{aligned} \tag{10.13}$$

where

$$\begin{aligned} p_{r,y-r}(n) &= \binom{y}{r}\binom{n-y}{\tfrac{1}{2}(n-1)-r} \Big/ \binom{n}{\tfrac{1}{2}(n-1)} \\ &= \binom{y}{r}\{\tfrac{1}{2}(n-1)\}^{(r)}\{\tfrac{1}{2}(n+1)\}^{(y-r)}/n^{(y)}. \end{aligned} \tag{10.14}$$

Of course, symmetry does not hold if n is odd, i.e. $p_{r,y-r}(n) \neq p_{y-r,r}(n)$.

The number of tests is not fixed, but depends on the sizes of remaining sets through all stages. It is, however, certain that if $n = 2^k + r\ (0 < r < 2^k)$ the number of tests must be either k or $(k+1)$. Calculations of $P(y, n)$ are not easy for $n \neq 2^k$. We can however determine values of $P(1, n)$.

We first note that if $n = 3$, then with probability $\tfrac{1}{3}$(or $\tfrac{2}{3}$), the NC item will be in the first (or second) group of a (1, 2) split, and with probability ϕ the group that contains the NC item will survive. A further (1, 1) split will be needed only if the NC is in the second group of two, and with chance ϕ it will finally remain. Hence

$$P(1,3) = \tfrac{1}{3}\phi + \tfrac{2}{3}\phi^2 = \frac{\phi}{3}(1 + 2\phi).$$

Generally, for m even,

$$P(1, m) = \phi P(1, \tfrac{1}{2}m), \tag{10.15a}$$

and, for m odd

$$\begin{aligned} P(1, m) &= \frac{m-1}{2m}\phi P\left(1, \frac{m-1}{2}\right) + \frac{m+1}{2m}\phi P\left(1, \frac{m+1}{2}\right) \\ &= \frac{\phi}{2m}\left\{(m-1)P\left(1, \frac{m-1}{2}\right) + (m+1)P\left(1, \frac{m+1}{2}\right)\right\}. \end{aligned} \tag{10.15b}$$

By induction we have

$$P(1, 2^k + r) = (2^k + r)^{-1}\phi^k(2^k - r + 2r\phi) \qquad \text{(for } r \text{ odd);} \tag{10.16a}$$

$$P(1, 2m) = \phi P(1, m); \tag{10.16b}$$

and

$$P(2, n) = \phi^{-1}P(1, n) \qquad \text{for } n \geqslant 2. \tag{10.16c}$$

PART THREE

Miscellaneous

CHAPTER 11

Estimation of error probabilities

11.1 Introduction

The problem of estimating the error probabilities p and p', and the related problems of testing hypotheses of perfect inspection ($p = 1$) and/or $p' = 0$) are among the most challenging in the field of attribute sampling. It is not possible to solve any of these problems unless some at least of the items are inspected more than once. In our discussion we use repeated inspection in various forms. If we can identify each inspection of the same item, then any disagreement between decisions at different inspections provids clear evidence that inspection is not error-free. Whether this reflects values of $p \neq 1$ or $p' \neq 0$, and what the values of p and/or p' may be, gives rise to much more difficult questions.

A variety of methods of estimation will be suggested in this chapter. The parameters of interest are the probabilities of correct (incorrect) classification of an NC (C) item, denoted by the symbols $p\,(p')$. The proportion of NC items, ω, may also be of interest.

In the next section, we will use maximum likelihood methods to estimate p, assuming that $p' = 0$ (or is negligible). In section 11.3, estimation by moments is discussed. Section 11.4 deals with situations wherein another measuring device (S) is introduced in addition to the test device (T), and both moment and maximum likelihood estimation are studied. Section 11.5 is on moment estimation for multitype nonconformities (introduced in Chapter 5). The last section is on hypothesis testing with $H_0{:}p = 1$ (assuming $p' = 0$).

11.2 Estimation with p' negligible

Suppose a sample of size n is chosen at random from lots of size N. The unknown number of NC items in this sample is denoted by y. We also assume that all items classified as NC must be NC, in other words, $p' = 0$. We will inspect the whole sample m times, and, consequently, each item may be declared NC several times, from zero to m. Let N_j denote the number of items which are declared NC just j times in the m trials. Therefore, we have $N_0 + N_1 + \cdots + N_m = n$. If inspection is perfect, $p = 1$, then $N_0 = n - y$, $N_m = y$. That means that we now have records of individual inspections, not just of the total number of NC decisions, which is $T = \sum_{j=1}^{m} jN_j$.

Maximum likelihood method

The likelihood function, conditional on y, is

$$L(y) = \binom{y}{y-n+N_0, N_1, \ldots, N_m}(1-p)^{m(y-n+N_0)} \times \prod_{j=1}^{m}\left\{\binom{m}{j}p^j(1-p)^{m-j}\right\}^{N_j}$$

$$= \binom{y}{y-n+N_0, N_1, \ldots, N_m} \times \left\{\prod_{j=1}^{m}\binom{m}{j}\right\}p^{\Sigma_1^m jN_j} \times (1-p)^{my-\Sigma_1^m jN_j}\Bigg\}, \qquad (11.1)$$

with $(n - N_0 \leqslant y \leqslant n)$. Note that $(N_0, \sum_{j=1}^{m} jN_j)$ is a sufficient statistic for (y, p). If y were known, the maximum likelihood estimator of p would be

$$\hat{p}(y) = (my)^{-1}\sum_{j=1}^{m} jN_j = (my)^{-1}T \qquad (11.2)$$

a function of y. From (11.1) and (11.2) the conditionally maximized log likelihood with respect to p would be

$$\log \hat{L}(y) = K + \sum_{i=1}^{n-N_0+1} \log(y-i) - (my-T)\log(my-T) - my\log(my), \qquad (11.3)$$

where K does not depend on y. From (11.3) we could find $\hat{y}$, which maximizes $\log \hat{L}(y)$ within the range $n \geqslant y \geqslant n - N_0$. Although $\hat{y}$ is not of direct interest, through it we can find the value of the maximum likelihood estimator, $\hat{p}(\hat{y})$, of p. Also, from $\hat{y}$ we can derive maximum likelihood estimators of ω (for an 'infinite' sized lot) or D (the number of NC items in a lot of size N). These are

$$\hat{\omega} = \hat{y}/n; \qquad \hat{D} = \text{integer part of } \hat{y}(N+1)/n.$$

A detailed discussion of the application of maximum likelihood methods to problems of this type is given by Lindsay (1985). In particular, he discusses situations in which $\hat{y} = n$. Lindsay and Roeder (1987) consider general problems arising in maximum likelihood estimation of integer-valued parameters. Dahiya (1981) and Olkin, Petkau and Zidek (1981), also discuss this problem; the former suggests a graphical method of solution. Estimation of ω from testing groups of individuals, with repeat group (*not* individual) test when a positive result is obtained is described by Gastwirth and Hammick (1989) and Hammick and Gastwirth (1990).

Moment method

We regard the N_j's, as observed values of independent binomial (m, p) variables truncated by omitting zero values (i.e. only for the $\sum_{j=1}^{m} N_j$ items declared NC at least once, which, under the assumption $p' = 0$, must indeed be NC items). Each of them has expected value $mp\{1-(1-p)^m\}^{-1}$, which, equating the sample means, gives the equation for the estimation ($\tilde{p}$).

$$m\tilde{p}\{1-(1-\tilde{p})^m\}^{-1} = (n - N_0)^{-1}T \tag{11.4}$$

= average number of times an item is declared NC, given that it is so declared at least once.

This estimator is, in fact, the same as the conditional maximum likelihood estimator of p, given N_0.

The equation (11.4) can be rewritten as

$$\tilde{p} = m^{-1}(n - N_0)^{-1}T\{1-(1-\tilde{p})^m\}, \tag{11.5}$$

which can be solved by iteration, although it can be easily solved directly for small of m, ($m \leqslant 3$, say). We note that (for infinite lot size*)

$$\operatorname{var}(T) = mnp(1-p)\omega + m^2p^2\omega(1-\omega);$$

where ω = proportion NC in the population and

$$\operatorname{var}(\tilde{p}) \doteq m^{-2}(n - N_0)^{-2}\{1-(1-\tilde{p})^m\}^2 \operatorname{var}(T).$$

With $p = \tilde{p}$,

$$E\left[\sum_{j=1}^{m} N_j = n - N_0 \,|\, y, \tilde{p}\right] = y\{1-(1-\tilde{p})\}^m,$$

*For finite lot size, the last term is multiplied by [(lot size) − n]/[(lot size) − 1].

which gives the following estimator of y, obtained by replacing expected values by observed values:

$$\tilde{y} = \min(n, [(n - N_0)\{1 - (1 - \tilde{p})^m\}^{-1}]) = \min(n, [(m\tilde{p})^{-1}T]), \tag{11.6}$$

where $[x]$ denotes 'nearest integer to x'.

A numerical example

Suppose that tests of $n = 50$ items, repeated $m = 3$ times each, give

$$N_0 = 43, \qquad N_1 = 1; \qquad N_2 = 1; \qquad N_3 = 5.$$

Then

$$n - N_0 = 7$$

$$T = \sum_{j=1}^{3} jN_j = 1 + 2 + 15 = 18.$$

We have (from (11.6))

$$\tilde{p}\{3\tilde{p} - 3\tilde{p}^2 + \tilde{p}^3\}^{-1} = (3 \times 7)^{-1}18 = \tfrac{6}{7},$$

whence $\tilde{p}^2 - 3\tilde{p} + \frac{11}{6} = 0$ leading to

$$\tilde{p} = 0{\cdot}8545$$

and

$$\tilde{y} = \min\left(50, \left[\frac{18}{3 \times 0{\cdot}8545}\right]\right) = 7.$$

Note that, whatever the value of n ($\geqslant 7$), values of $\tilde{p}$ and $\tilde{y}$ would be the same.

11.3 General estimation by moments

We are mainly interested in estimating p and p' with unknown ω, which is the proportion of NC items in the lot, as a nuisance parameter. As we have mentioned, it is impossible to estimate p, p' and ω by inspecting each item just once, although the probability of declaring a randomly selected item NC, which equals $\omega p + (1 + \omega)p'$, can be estimated in this way.

Suppose the sample of size n is inspected m times. Let

M_i = number of times of ith item is declared NC;
N_j = number of items declared NC j times;

with

$$i, N_j = 1, \ldots, n, \qquad \text{and} \qquad j, M_i = 0, \ldots, m.$$

Note that $\sum_{i=1}^{n} M_i = T = \sum_{j=1}^{m} jN_j$,

Estimation

Clearly the distribution of M_i is

$$\Pr[M_i = m_i] = \omega\binom{m}{m_i}p^{m_i}(1-p)^{m-m_i} + (1-\omega)\binom{m}{m_i}p'^{m_i}(1-p')^{m-m_i}$$
$$(m_i = 0, \ldots, m). \quad (11.7)$$

The rth factorial moment of each M_i is

$$\mu_{(r)} = E[M_i^{(r)}] = E[M_i(M_i-1)\cdots(M_i-r+1)] = m^{(r)}\{\omega p^r + (1-\omega)p'^r\}, \quad (11.8)$$

and equating the first three factorial moments to the corresponding sample values gives

$$\tilde{\omega}\tilde{p}^r + (1-\tilde{\omega})\tilde{p}'^r = F_r \qquad (r = 1, 2, 3) \quad (11.9)$$

where

$$F_r = (m^{(r)})^{-1} n^{-1} \sum_{i=1}^{n} M_i^{(r)}.$$

Solutions $\tilde{p}$, $\tilde{p}'$, and $\tilde{\omega}$ of (11.9) are given by Jones (1933) as follows:

1. $\tilde{p}, \tilde{p}'$ are roots (in θ) of the equation

$$\theta^2 - A\theta + AF_1 - F_2 = 0 \quad (11.10)$$

where $A = (F_3 - F_1F_2)/(F_2 - F_1^2)$. (Note that $AF_1 - F_2 = (F_1F_3 - F_2^2)/(F_2 - F_1^2)$.)

2.

$$\tilde{\omega} = (F_1 - \tilde{p}')/(\tilde{p} - \tilde{p}'). \quad (11.11)$$

Equation (11.10) has two solutions. We will take the greater one as

$\tilde{p}$ with the common assumption, $p > p'$, in mind (although $p > p'$ does not necessarily imply $\tilde{p} > \tilde{p}'$).

Example

We will use the fact that

$$\sum_{i=1}^{n} M_i^{(r)} = \sum_{j=1}^{m} N_j j^{(r)}, \tag{11.12}$$

where N_j is the number of items declared NC on just j occasions among the m rounds of examination.

Suppose we have (as on p. 158) $n = 50$, $m = 3$; $N_0 = 43$; $N_1 = 1$, $N_2 = 1$, $N_3 = 5$. Then

$$F_r = (3^{(r)} \cdot 50)^{-1}(0^{(r)} \cdot 43 + 1^{(r)} \cdot 1 + 2^{(r)} \cdot 1 + 3^{(r)} \cdot 5),$$

whence

$$F_1 = \tfrac{1}{150}(1 + 2 + 15) = \tfrac{3}{25};$$
$$F_2 = \tfrac{1}{300}(2 + 30) = \tfrac{8}{75};$$
$$F_3 = \tfrac{1}{300} \cdot 30 = \tfrac{1}{10}.$$

So $A = 327/346 = 0{\cdot}94508671$ and $\tilde{p}, \tilde{p}'$ are roots of $\theta^2 - 0{\cdot}94519\theta + 0{\cdot}00674 = 0$. We find $\tilde{p} = 0{\cdot}9379$; $\tilde{p}' = 0{\cdot}0072$; $\tilde{\omega} = 0{\cdot}1212$.

In this case $\tilde{p}, \tilde{p}'$ and $\tilde{\omega}$ are all between 0 and 1.

Variances

The following asymptotic formulas for the variances of $\tilde{p}$, $\tilde{p}'$ and $\tilde{\omega}$ are given by Blischke (1962, 1964):

$$\operatorname{var}(\tilde{p}) \doteq \frac{p(1-p)}{\omega nm} + \frac{2(4B_2 + B_2')}{nm(m-1)C_2} + \frac{6(B_3 + B_3')}{nm(m-1)(m-2)C_4}, \tag{11.13a}$$

$$\operatorname{var}(\tilde{p}') \doteq \frac{p'(1-p')}{(1-\omega)nm} + \frac{2(B_2 + 4B_2')}{nm(m-1)C_2'} + \frac{6(B_3 + B_3')}{nm(m-1)(m-2)C_4'}, \tag{11.13b}$$

$$\operatorname{var}(\tilde{\omega}) \doteq \frac{\omega(1-\omega)}{n} + \frac{18(B_2 + B_2')}{nm(m-1)(p-p')^4} + \frac{24(B_3 + B_3')}{nm(m-1)(m-2)(p-p')^6}. \tag{11.13c}$$

where

$$B_h = \omega p^h(1-p)^h; \quad B'_h = (1-\omega)p'^h(1-p')^h;$$
$$C_h = \omega^2(p-p')^h; \quad C'_h = (1-\omega)^2(p-p')^h.$$

Taking the first term of each of the above formulas as an approximation when m is large, we have

$$\text{var}(\tilde{p}) \doteq \frac{p(1-p)}{\omega mn};$$
$$\text{var}(\tilde{p}') \doteq \frac{p'(1-p')}{(1-\omega)mn}; \qquad (11.14)$$
$$\text{var}(\tilde{\omega}) \doteq \frac{\omega(1-\omega)}{n}.$$

Note the three denominators:

1. $\omega mn = E$[the number of inspections of NC items];
2. $(1-\omega)mn = E$[the number of inspections of C items];
3. $n =$ the number of items in the sample.

With the numerical values used in the previous example, and inserting the values $\tilde{p}, \tilde{p}', \tilde{\omega}$ for p, p', ω, respectively, we find

$$\text{var}(\tilde{p}) \doteq 0{\cdot}004133, \qquad \text{var}(\tilde{p}') \doteq 0{\cdot}0000591, \qquad \text{var}(\tilde{\omega}) \doteq 0{\cdot}002170.$$

The last two unused terms on the right-hand sides of (11.13a)–(11.13c) are very small, even for $m = 3$, when they are

0·000885 and 0·000044 for $\text{var}(\tilde{p})$;

0·000004 and 0·000001 for $\text{var}(\tilde{p}')$;

0·000036 and 0·000003 for $\text{var}(\tilde{\omega})$.

The approximation is quite good in this case.

Confidence intervals

The asymptotic normality of the estimators (shown by Blischke (1962)) gives approximate $100(1-\alpha)\%$ confidence regions for ω from the relation

$$\frac{n(\tilde{\omega}-\omega)^2}{\omega(1-\omega)} < \lambda^2_{\alpha/2},$$

where $\Phi(-\lambda_{\alpha/2})=\frac{1}{2}\alpha$ and $\Phi(y)=(\sqrt{2\pi})^{-1}\int_{-\infty}^{y} e^{-u^2/2}\,du$. Taking $\alpha=0{\cdot}05$, so that $\lambda^2_{\alpha/2}=\lambda^2_{0\cdot025}=3{\cdot}8416$, the approximate 95% region for ω is

$$50(0{\cdot}1212-\omega)^2<3{\cdot}8416\omega(1-\omega), \qquad \text{i.e.} \quad 0{\cdot}057<\omega<0{\cdot}0240.$$

The corresponding region for p (and similarly for p') would be

$$\frac{\omega mn(\tilde{p}-p)^2}{p(1-p)}<\lambda^2_{\alpha/2},$$

which is not usable unless the unknown ω is replaced by $\tilde{\omega}$, which would give an asymptotically correct region.

With $\tilde{\omega}=0{\cdot}1212$, and $\alpha=0{\cdot}05$, as before, we get the (approximate) 95% confidence intervals:

$$0{\cdot}7350<p<0{\cdot}9880;\ 0{\cdot}00023<p'<0{\cdot}0650.$$

11.4 Estimation with two inspection devices

Introduction and notation

Suppose we inspect a random sample of size n from an 'infinite' lot with proportion ω of NC items. The probability of classifying an item as NC correctly (incorrectly) is $p(p')$. Let Z be the number of declared NC items from the sample by the inspection. We know that

$$Z \sim \text{Bin}\,(n,\theta),$$

with $\theta=\omega p+(1-\omega)p'$. We also know that it is impossible to estimate p and p' by using one kind of inspection on a single sample just once. Hochberg and Tenenbein (1983) (see also Tenenbein, 1970) propose the introduction of another inspection device, S say, with known values of p and p' which are denoted p_S and p'_S, and call the original device T with the unknown values of p and p' denoted p_T and p'_T. We assume that $p_S \gg p'_S$ and $p_T \gg p'_T$ when necessary. We also assume that S and T classifications are independent, given the true status of the items.

Estimating p_T and p'_T is our main interest, with the unknown proportion ω as a nuisance parameter.

We will use the following symbols to describe data from the two kinds of inspections S and T:

1. n_S items using S device alone, with Z_S judged as NC;
2. n_T items using T device alone, with Z_T judged as NC;

3. n items using both S and T, with results shown below:

	Number of NC by T	Number of C by T
Number of NC by S	Z_{11}	Z_{10}
Number of C by S	Z_{01}	Z_{00}

Clearly, $Z_{11} + Z_{10} + Z_{01} + Z_{00} = n$, and let $N \equiv n_S + n_T + n$ denote the total number of observations. From our assumptions, we have

$$Z_S, Z_T \text{ and } \mathbf{Z} \equiv \begin{pmatrix} Z_{11} & Z_{10} \\ Z_{01} & Z_{00} \end{pmatrix} \text{ are mutually independent.} \quad (11.15)$$

Define

$$\theta_S = p_S\omega + p'_S(1 - \omega); \quad (11.16)$$

$$\theta_T = p_T\omega + p'_T(1 - \omega); \quad (11.17)$$

$$\phi = p_S p_T \omega + p'_S p'_T (1 - \omega). \quad (11.18)$$

We have

$$Z_S \sim \text{Bin}(n_S, \theta_S); \quad (11.19)$$

$$Z_T \sim \text{Bin}(n_T, \theta_T). \quad (11.20)$$

Also, by our assumptions

$$\mathbf{Z} \sim \text{Multinomial}\left(n; \begin{pmatrix} \phi & \theta_S - \phi \\ \theta_T - \phi & 1 - \theta_S - \theta_T + \phi \end{pmatrix}\right). \quad (11.21)$$

Moment estimation

From (11.16)–(11.18),

$$\omega = (\theta_S - p'_S)/(p_S - p'_S); \quad (11.22)$$

$$p_T = (\phi - p'_S\theta_T)/(\theta_S - p'_S); \quad (11.23)$$

$$p'_T = (p_S\theta_T - \phi)/(p_S - \theta_S). \quad (11.24)$$

Now we define unbiased estimators $\tilde{\theta}_S, \tilde{\theta}_T$ and $\tilde{\phi}$ of θ_S, θ_T and ϕ, respectively, as follows:

$$(n_S + n)\tilde{\theta}_S = Z_S + Z_{10} + Z_{11} \sim \text{Bin}(n_S + n, \theta_S); \quad (11.25)$$

$$(n_T + n)\tilde{\theta}_T = Z_T + Z_{01} + Z_{11} \sim \text{Bin}(n_T + n, \theta_T); \quad (11.26)$$

$$n\tilde{\phi} = Z_{11} \sim \text{Bin}(n, \phi). \quad (11.27)$$

Replacing θ_S, θ_T and ϕ in (11.22)–(11.24) by $\tilde{\theta}_S, \tilde{\theta}_T$ and $\tilde{\phi}$, we have estimators $\tilde{\omega}, \tilde{p}_T$ and $\tilde{p}'_T$ of ω, p_T and p'_T:

$$\tilde{\omega} = (\tilde{\theta}_S - p'_S)/(p_S - p'_S); \tag{11.28}$$

$$\tilde{p}_T = (\tilde{\phi} - p'_S\tilde{\theta}_T)/(\tilde{\theta}_S - p'_S); \tag{11.29}$$

$$\tilde{p}'_T = (p_S\tilde{\theta}_T - \tilde{\phi})/(p_S - \tilde{\theta}_S). \tag{11.30}$$

Among these estimators, $\tilde{\omega}$ is unbiased while $\tilde{p}_T$ and $\tilde{p}'_T$ are biased, although the biases may not be large for adequate sample sizes.

The variance–covariance matrix of the random variables in (11.25)–(11.27) is

$$\mathbf{Var}((n_S + n)\tilde{\theta}_S, (n_T + n)\tilde{\theta}_T, n\tilde{\phi})$$
$$= \begin{bmatrix} (n_S + n)\theta_S(1 - \theta_S) & n(\phi - \theta_S - \theta_T) & n\phi(1 - \theta_S) \\ n(\phi - \theta_S\theta_T) & (n_T + n)\theta_T(1 - \theta_T) & n\phi(1 - \theta_T) \\ n\phi(1 - \theta_S) & n\phi(1 - \theta_T) & n\phi(1 - \theta) \end{bmatrix}.$$

Therefore

$$\operatorname{var}(\tilde{\omega}) = (n_S + n)^{-1}(p_S - p'_S)^{-2}\theta_S(1 - \theta_S), \tag{11.31}$$

and, using the method of statistical differentials (e.g. Johnson and Kotz (1985a) Chapter 1, section 7.5) we obtain, after some algebra, the approximation

$$\begin{aligned}\operatorname{var}(\tilde{p}_T) \doteq p_T^2\omega^{-2}(p_S - p'_S)^{-2}[\{&n^{-1}\phi(1 - \phi) - 2(n_T + n)^{-1} \\ &\times p'_S\phi(1 - \theta_T) + (n_T + n)^{-1}p_S'^2\theta_T(1 - \theta_T)\}p^{-2} \\ &- 2(n_S + n)^{-1}\{\phi(1 - \theta_S) - n(n_T + n)^{-1}p'_S(\phi - \theta_S\theta_T)\}p_T^{-1} \\ &+ (n_S + n)^{-1}\theta_S(1 - \theta_S)].\end{aligned} \tag{11.32}$$

A similar approximation for $\operatorname{var}(\tilde{p}'_T)$ can be obtained from (11.32) by noticing the symmetry between p and p' (simply replace p_T by p'_T and ω by $(1 - \omega)$, and interchange p_S and p'_S).

An approximation to the bias of $\tilde{p}_T$ is given by

$$E[\tilde{p}_T] - p_T \doteq p_T\left\{\frac{\operatorname{var}(\tilde{\theta}_S)}{(\theta_S - p'_S)^2} - \frac{\operatorname{cov}(\tilde{\theta}_S, \tilde{\phi} - p'_S\tilde{\theta}_T)}{(\theta_S - p'_S)(\phi - p'_S\theta_T)}\right\}, \tag{11.33}$$

which gives a proportional bias [i.e. $100 \times \text{bias}/p_T\%$]

$$\frac{100\{n_T p'_S(1 - \theta_S) + n(1 - p'_S)\theta_S\}(\phi - \theta_S\theta_T)}{(n_S + n)(n_T + n)\omega^2(p_S - p'_S)^2 p_T}\%. \tag{11.34}$$

This is positive and, since $p'_T < p_T$, less than

$$\frac{100G(1-\omega)}{(n_S+n)\omega^2(p_S-p'_S)^2}\%, \tag{11.35}$$

where

$$G = \frac{n_T}{n_T+n}p'_S(1-\theta_S) + \frac{n}{n_T+n}(1-p'_S)\theta_S. \tag{11.36}$$

Note that G is between $p'_S(1-\theta_S)$ and $(1-p'_S)\theta_S$.

Example 1

With $p_S = 0{\cdot}9$, $p'_S = 0{\cdot}1$ and $\omega = 0{\cdot}1$ we have $\theta_S = 0{\cdot}18$ and

$$G = (n_T+n)^{-1}(0{\cdot}082n_T + 0{\cdot}162n).$$

Therefore $0{\cdot}082 \leqslant G \leqslant 0{\cdot}162$ and the approximate proportional bias of $\tilde{p}_T$ is between 0 and $1406{\cdot}25G(n_S+n)^{-1}\%$. Note that the upper limit is less than $227{\cdot}8(n_S+n)^{-1}\%$, so if $n_S + n > 100$ the approximate proportional bias is less than 2.28%.

Maximum likelihood estimation

We first estimate θ_S, θ_T and ϕ, and then obtain the estimators of p, p' and ω from the relations:

$$p'_S = (\theta_S - p_S\omega)(1-\omega)^{-1},$$

$$p_T = \frac{\phi(1-\omega) - (\theta_S - p_S\omega)\theta_T}{(p_S-\theta_S)\omega},$$

$$p'_T = p_S\theta_S - \phi p_S - \theta_S.$$

The likelihood function of Z_S, Z_T, and $\mathbf{Z}$ is

$$\binom{n_S}{Z_S}\binom{n_T}{Z_T}\binom{n}{Z_{11}, Z_{10}, Z_{01}, Z_{00}}\theta_S^{Z_S}(1-\theta_S)^{n_S-Z_S}\theta_T^{Z_T}$$
$$\times(1-\theta_T)^{n_T-Z_T}\phi^{Z_{11}}(\theta_S-\phi)^{Z_{10}}(\theta_T-\phi)^{Z_{01}}(1-\theta_S-\theta_T+\phi)^{Z_{00}}.$$

Taking partial derivatives and equating them to zero, we obtain equations for the maximum likelihood estimators (MLE)$\hat{\theta}_S$, $\hat{\theta}_T$, and $\hat{\phi}$:

$$\frac{Z_S}{\hat{\theta}_S} - \frac{n_S - Z_S}{1-\hat{\theta}_S} + \frac{Z_{10}}{\hat{\theta}_S - \hat{\phi}} - \frac{Z_{00}}{1-\hat{\theta}_S-\hat{\theta}_T+\hat{\phi}} = 0; \tag{11.37}$$

$$\frac{Z_\mathrm{T}}{\hat{\theta}_\mathrm{T}} - \frac{n_\mathrm{T} - Z_\mathrm{T}}{1 - \hat{\theta}_\mathrm{T}} + \frac{Z_{01}}{\hat{\theta}_\mathrm{T} - \hat{\phi}} - \frac{Z_{00}}{1 - \hat{\theta}_\mathrm{S} - \hat{\theta}_\mathrm{T} + \hat{\phi}} = 0; \tag{11.38}$$

$$\frac{Z_{11}}{\hat{\theta}} - \frac{Z_{10}}{\hat{\theta}_\mathrm{S} - \hat{\phi}} - \frac{Z_{01}}{\hat{\theta}_\mathrm{T} - \hat{\phi}} + \frac{Z_{00}}{1 - \hat{\theta}_\mathrm{S} - \hat{\theta}_\mathrm{T} + \hat{\phi}} = 0; \tag{11.39}$$

subject to $0 < \phi < \theta_\mathrm{S}, \theta_\mathrm{T} < 1$ and $\phi > \theta_\mathrm{S} + \theta_\mathrm{T} - 1$. With symbols

$$f(a,b) \equiv \frac{n(1-\theta_a)}{(\theta_b - \phi)(1 - \theta_b - \theta_a + \phi)};$$

$$g(a,b) \equiv \frac{n}{1 - \theta_b - \theta_a + \phi};$$

$$h(a,b) \equiv \frac{n_b}{\theta_b(1-\theta_b)} + f(a,b);$$

$$\kappa \equiv \frac{n\{\theta_\mathrm{S}\theta_\mathrm{T}(1 - \theta_\mathrm{S} - \theta_\mathrm{T}) + 2\theta_\mathrm{S}\theta_\mathrm{T}\phi - \phi^2\}}{\phi(\theta_\mathrm{S} - \phi)(\theta_\mathrm{T} - \phi)(1 - \theta_\mathrm{S} - \theta_\mathrm{T} + \phi)};$$

the information (inverse variance-covariance) matrix is

$$\mathbf{V}^{-1} = \begin{bmatrix} h(\mathrm{T},\mathrm{S}) & g(\mathrm{T},\mathrm{S}) & f(\mathrm{T},\mathrm{S}) \\ g(\mathrm{T},\mathrm{S}) & h(\mathrm{S},\mathrm{T}) & f(\mathrm{S},\mathrm{T}) \\ f(\mathrm{T},\mathrm{S}) & f(\mathrm{S},\mathrm{T}) & \kappa \end{bmatrix}, \tag{11.40}$$

and the determinant

$$|\mathbf{V}^{-1}| = \frac{n}{\phi(\theta_\mathrm{S} - \phi)(\theta_\mathrm{T} - \phi)(1 - \theta_\mathrm{S} - \theta_\mathrm{T} + \phi)} \times \left[\frac{n_\mathrm{S} n_\mathrm{T} \gamma}{\theta_\mathrm{S}\theta_\mathrm{T}(1-\theta_\mathrm{S})(1-\theta_\mathrm{T})} + n(n_\mathrm{S} + n_\mathrm{T} + n)\right],$$

where $\gamma = \theta_\mathrm{S}\theta_\mathrm{T}(1 - \theta_\mathrm{S} - \theta_\mathrm{T} + \phi) - \phi(\phi - \theta_\mathrm{S}\theta_\mathrm{T})$. From the asymptotic variance–covariance matrix $\mathbf{V}$ we obtain

$$\begin{aligned} \mathrm{var}(\hat{\theta}_\mathrm{S}) &\doteq \frac{n}{|\mathbf{V}^{-1}|}\left\{\frac{n_\mathrm{T}}{\theta_\mathrm{T}(1-\theta_\mathrm{T})}\left(\frac{1}{\phi} + \frac{1}{\theta_\mathrm{S} - \phi} + \frac{1}{\phi_\mathrm{T} - \phi} + \frac{1}{1 - \theta_\mathrm{S} - \theta_\mathrm{T} + \phi}\right)\right. \\ &\qquad \left. + n\left(\frac{1}{\theta_\mathrm{T} - \phi} + \frac{1}{1 - \theta_\mathrm{S} - \theta_\mathrm{T} + \phi}\right)\left(\frac{1}{\phi} + \frac{1}{\theta_\mathrm{S} - \phi}\right)\right\} \\ &\doteq \theta_\mathrm{S}(1-\theta_\mathrm{S})(\gamma n_\mathrm{S}^{-1} + \delta N^{-1})/(\gamma + \delta), \end{aligned} \tag{11.41}$$

where $N = n_S + n_T + n$ and $\delta = (nN/n_S n_T)\theta_S\theta_T(1-\theta_S)(1-\theta_T)$.

The MLE of ω is

$$\hat{\omega} = (p_S - p'_S)^{-1}(\hat{\theta}_S - p'_S). \tag{11.42}$$

The asymptotic efficiency of $\tilde{\omega}$ [see (11.28)] relative to $\hat{\omega}$ is the same as that of $\tilde{\theta}_S$ relative to $\hat{\theta}_S$, which is

$$100(n_S + n)(\gamma n_S^{-1} + \delta N^{-1})/(\gamma + \delta)\%. \tag{11.43}$$

Example 1 (continued)

Taking $p_S = p_T = 0{\cdot}9$, $p'_S = p'_T = 0{\cdot}1 = \omega$, as in Example 1, and $n_S = n_T = n\ (= (1/3)N)$ we find $\gamma = 0{\cdot}0184680$ and $\delta = 0{\cdot}0653573$, so (11.43) becomes

$$100(n_S + n)(0{\cdot}2203 n_S^{-1} + 0{\cdot}7797 N^{-1}) = 2(0{\cdot}2203 + 0{\cdot}2599) = 96{\cdot}04\%.$$

The asymptotic variance of the MLE $\hat{\phi}$ is

$$\begin{aligned}\operatorname{var}(\hat{\phi}) \doteq \frac{1}{n}[&\delta\phi(1-\phi) - \delta\phi^2 N^{-1}\{n_S\theta_S^{-1}(1-\theta_S) + n_T\theta_T^{-1}(1-\theta_T)\} \\ &+ \phi(\theta_S - \phi)(\theta_T - \phi)(1 - \theta_S - \theta_T + \phi)]/(\delta + \gamma).\end{aligned} \tag{11.44}$$

Since, for the moment estimator $\tilde{\phi}$,

$$\operatorname{var}(\tilde{\phi}) = n^{-1}\phi(1-\phi),$$

we can find the asymptotic efficiency of $\tilde{\phi}$ relative to $\hat{\phi}$. For the above example it is

$$\begin{aligned}&100 \times [0{\cdot}0653573 \times 0{\cdot}09\{0{\cdot}91 - \tfrac{2}{3} \times 0{\cdot}09 \times (0{\cdot}81)^{-1} \times 0{\cdot}82\} \\ &\qquad + 0{\cdot}09 \times 0{\cdot}09^2 \times 0{\cdot}73]/[0{\cdot}09 \times 0{\cdot}91(0{\cdot}0653573 + 0{\cdot}0184680)] \\ &= 100 \times \frac{0{\cdot}0037449 + 0{\cdot}0005322}{0{\cdot}0068653} = 62{\cdot}30\%.\end{aligned}$$

The low asymptotic efficiency of $\tilde{\phi}$ with respect to $\hat{\phi}$ reflects that the moment method does not use the information on values of θ_S and θ_T contained in the other $(n_S + n_T)$ observations.

Calculation of MLE

It is impossible to obtain explicit solutions of (11.37)–(11.39) for $\hat{\theta}_S$, $\hat{\theta}_T$, and $\hat{\phi}$, but numerical solutions can be obtained. An EM algorithm (e.g. Dempster *et al.* (1977)) could be constructed in the following way. If we tested the group of n_S by T and the group of n_T by S, we would have every item (total $n_T + n_S$) tested by both T and S, and that would give values similar to the Z_{ij}'s. Formally, let $Z_{ij(S)}$ ($Z_{ij(T)}$ ($i = j = 0, 1$) be the (unobserved) numbers of items with i, j decision combinations which would have been obtained if the $n_S(n_T)$ items, which were tested by S(T), had also been tested by T(S). Clearly,

$$Z_{10(S)} + Z_{11(S)} = Z_S \qquad \text{and} \qquad Z_{10(T)} + Z_{11(T)} = Z_T.$$

If values of these variables had been observed, the MLE's would have been as follows:

$$\text{For } \theta_S\text{:} (Z_S + Z_{10(T)} + Z_{11(T)} + Z_{10} + Z_{11})N^{-1}; \tag{11.45}$$

$$\text{For } \theta_T\text{:} (Z_{01(S)} + Z_{11(S)} + Z_{(T)} + Z_{01} + Z_{11})N^{-1}; \tag{11.46}$$

$$\text{For } \phi\text{:} (Z_{11(S)} + Z_{11(T)} + Z_{11})N^{-1}. \tag{11.47}$$

Note that

$$E[Z_{10(T)}|Z_T] = (n_T - Z_T)(\theta_S - \phi)(1 - \theta_T)^{-1};$$
$$E[Z_{11(T)}|Z_T] = Z_T\phi\theta_T^{-1};$$
$$E[Z_{01(S)}|Z_S] = (n_S - Z_S)(\theta_T - \phi)(1 - \theta_S)^{-1};$$
$$E[Z_{11(S)}|Z_S] = Z_S\phi\theta_S^{-1}.$$

Replacing $Z_{i,j(\cdot)}$ in (11.45)–(11.47) by the above corresponding conditional expectations leads to iteration formulas

$$\theta_S^{(v+1)} = N^{-1}\left[Z_S + \frac{(n_T - Z_T)(\theta_S^{(v)} - \phi^{(v)})}{1 - \theta_T^{(v)}} + \frac{Z_T\phi^{(v)}}{\theta_T^{(v)}} + Z_{10} + Z_{11}\right], \tag{11.48}$$

$$\theta_T^{(v+1)} = N^{-1}\left[\frac{(n_S - Z_S)(\theta_T^{(v)} - \phi^{(v)})}{1 - \theta_S^{(v)}} + \frac{Z_S\phi^{(v)}}{\theta_S^{(v)}} + Z_{10} + Z_{11}\right], \tag{11.49}$$

$$\phi^{(v+1)} = N^{-1}\left[\left(\frac{Z_T}{\theta_T^{(v)}} + \frac{Z_S}{\theta_S^{(v)}}\right)\phi^{(v)} + Z_{11}\right]. \tag{11.50}$$

Johnson and Kotz (1988a) give an example.

Remarks

Maximum likelihood estimation gives better results than the methods of moments. Of course, the moment estimators could be used as the initial values of iteration for MLE. Because of our assumptions, the correctness of values of p_S and p'_S is very important. For instance, large values of either of them may lead to greater negative bias in estimates of ω. If, in addition, ω is known, then so is θ_S, and it is not necessary to use the S device alone, i.e. we can take $n_S = 0$.

11.5 Moment estimation on multitype nonconformities

Assumptions and notation

Suppose there are k NC types in a lot of size N. Each item has an NC pattern $(\mathbf{g}) = (g_1, \ldots, g_k)$ with $g_i = 0$ or 1, where $g_i = 1$ indicates the presence of NC type i, and $g_i = 0$ means the absence of the ith type of nonconformity (cf. section 5.4). Obviously, there are, in all, 2^k different patterns. Let $D_{\mathbf{g}}$ be the number of items with NC pattern $(\mathbf{g})$ in the lot. Suppose a random sample of size n is chosen and inspected, and all decisions regarding the presence or absence of each type or joint set of types are mutually independent. Let p_j denote the probability of correct detection of the jth NC type and p'_j the probability of false detection of NC type j. In the sample, we use $Z_{\mathbf{g}}$ to denote the number of items declared as having NC pattern $(\mathbf{g})$ and $Y_{\mathbf{g}}$ the number of items which actually have NC pattern $(\mathbf{g})$. Clearly

$$\sum_{g_1=0}^{1} \cdots \sum_{g_k=0}^{1} D_{\mathbf{g}} = \sum_{\mathbf{g}} D_{\mathbf{g}} = N, \tag{11.51}$$

and

$$\sum_{\mathbf{g}} Y_{\mathbf{g}} = n. \tag{11.52}$$

The probability that an item truly having NC pattern $(\mathbf{f})$ will be judged to have NC pattern $(\mathbf{g})$ by inspection is

$$p_{\mathbf{g}|\mathbf{f}} = \prod_{i=1}^{k} p_i^{f_i g_i} p_i'^{(1-f_i)g_i} (1-p_i)^{f_i(1-g_i)} (1-p'_i)^{(1-f_i)(1-g_i)}. \tag{11.53}$$

We assume $p_j > p'_j$ for all j. This assumption is reasonable; without it the inspection would be totally unreliable.

Moment estimators

Our aim is to estimate the $D_{\mathbf{g}}$'s through observed $Z_{\mathbf{g}}$'s. We know that

$$\mathbf{Z} = \{Z_{\mathbf{g}}\} \sim *\cdots*_{\mathbf{f}} \text{ Multinomial}(Y_{\mathbf{g}}, \{p_{\mathbf{g}|\mathbf{f}}\}) \bigwedge_{\mathbf{Y}} \text{Mult Hypg}(n; \mathbf{D}; N).$$

(where $\mathbf{D} = \{D_{\mathbf{g}}\}$)
which leads to

$$E[Z_{\mathbf{g}}] = E[Z_{\mathbf{g}} | \{Y_{\mathbf{f}}\}] = \sum_{(\mathbf{f})} p_{\mathbf{g}|\mathbf{f}} E[Y_{\mathbf{f}}] = n N^{-1} \sum_{(\mathbf{f})} p_{\mathbf{g}|\mathbf{f}}(D_{\mathbf{f}}). \tag{11.54}$$

Let $\tilde{D}_{\mathbf{f}}$ be the moment estimator of $D_{\mathbf{f}}$. There are 2^k of these estimators, corresponding to the 2^k different patterns. They satisfy the 2^k equations

$$Z_{\mathbf{g}} = n N^{-1} \sum_{(\mathbf{f})} p_{\mathbf{g}|\mathbf{f}} \tilde{D}_{\mathbf{f}}, \tag{10.55}$$

or, in matrix form,

$$\mathbf{Z} = n N^{-1} \mathbf{P}\tilde{\mathbf{D}}, \tag{11.55$'$}$$

where $\mathbf{Z} = \{Z_{\mathbf{g}}\}$ and $\tilde{\mathbf{D}} = \{\tilde{D}_{\mathbf{g}}\}$ are two 2^k vectors, and $\mathbf{P}$ is a $2^k \times 2^k$ matrix with $p_{\mathbf{g}|\mathbf{f}}$ as its elements (rows correspond to the 2^k (**g**) patterns and columns to the 2^k (**h**) patterns). We therefore have

$$\tilde{\mathbf{D}} = N n^{-1} \mathbf{P}^{-1} \mathbf{Z}, \tag{11.56}$$

where $\mathbf{P}^{-1}$ is the left inverse of $\mathbf{P}$. It is clear that $\tilde{\mathbf{D}}$ is an unbiased estimator $\mathbf{D}$, in fact, $E[\tilde{D}_{\mathbf{g}}] = D_{\mathbf{g}}$ for all (**g**).

Note that since $\sum_{\mathbf{g}} \tilde{D}_{\mathbf{g}} = N$, it is possible that the $\tilde{D}_{\mathbf{g}}$'s might be negative as well as rarely exceeding N.

The $k = 1$ case. For a single NC type case, i.e. $k = 1$, we have $D_1 = N - D_0$ and

$$E[Z_1] = N^{-1} n\{D_1 p_1 + (N - D_1)p'_1\} = N^{-1} n\{D_1(p_1 - p'_1) + N p'_1\},$$

whence

$$\tilde{D}_1 = N(p_1 - p'_1)^{-1}(n^{-1} Z_1 - p'_1). \tag{11.57}$$

Therefore, (by assuming $p_1 > p'_1$),

$$\tilde{D}_1 \begin{cases} < 0 & \text{if } Z_1 < np'_1; \\ = N & \text{if } Z_1 = np'_1; \\ > N & \text{if } Z_1 > np'_1. \end{cases} \tag{11.58}$$

A modified, but not unbiased, estimator avoiding 'impossible' values

(<0 or $>N$) is

$$\tilde{D}_1^* = \begin{cases} 0 & \text{if } \tilde{D}_1 < 0; \\ D_1 & \text{if } 0 \leqslant \tilde{D}_1 \leqslant N; \\ N. & \text{if } N < \tilde{D}_1. \end{cases} \tag{11.59}$$

Table 1 of Johnson and Kotz (1984b) gives values of

$$\Pr[\tilde{D}_1 < 0] \qquad \text{and} \qquad \text{Bias}(\tilde{D}_1^*) = E[\tilde{D}_1^*] - D_1$$

for various sets of parameters (they are both zero when $p_1' = 0$).

Distributions

Let $W_{\mathbf{g}|\mathbf{f}}$ be the number of items which actually belong to the group with pattern ($\mathbf{f}$) but are declared to have pattern ($\mathbf{g}$). There are 2^k of those $W_{\mathbf{g}|\mathbf{f}}$'s, given each pattern ($\mathbf{f}$). Conditional on $\mathbf{Y} = \{Y_{\mathbf{f}}\}$, the joint distribution of the 2^k $W_{\mathbf{g}|\mathbf{f}}$'s is multinomial with parameters ($Y_{\mathbf{f}}$; $\{p_{\mathbf{p}|\mathbf{f}}\}$). $W_{\mathbf{g}|\mathbf{f}}$ and $W_{\mathbf{g}^*|\mathbf{f}^*}$ are mutually independent and binomially distributed with parameters ($Y_{\mathbf{f}}$; $p_{\mathbf{g}|\mathbf{f}}$) and ($Y_{\mathbf{f}^*}$; $p_{\mathbf{g}^*|\mathbf{f}^*}$), respectively, for different ($\mathbf{f}$) and ($\mathbf{f}^*$) and any ($\mathbf{g}$) and ($\mathbf{g}^*$) (not necessarily different). Because

$$Z_{\mathbf{g}} = \sum_{\mathbf{f}} W_{\mathbf{g}|\mathbf{f}}, \tag{11.60}$$

the joint distribution of $\mathbf{Z}$, conditioned on $\mathbf{Y}$, is

$$\mathbf{Z}|\mathbf{Y} \sim * \cdots *_{(\mathbf{f})} \text{Multinomial}(Y_{\mathbf{f}}; \{p_{\mathbf{g}|\mathbf{f}}\}). \tag{11.61}$$

(Of course, $\sum_{\mathbf{g}} W_{\mathbf{g}|\mathbf{f}} = Y_{\mathbf{f}}$.)

The overall distribution of $\mathbf{Z}$ is

$$\mathbf{Z} \sim * \cdots *_{(\mathbf{g}|\mathbf{f})} \text{Multinomial}(Y_{\mathbf{f}}; p_{\mathbf{g}|\mathbf{f}}\}) \bigwedge_{\mathbf{Y}} \text{Multi Hypg}(n; \mathbf{D}; N), \tag{11.62}$$

or

$$\Pr[\mathbf{Y} = \mathbf{y}] = \binom{N}{n}^{-1} \prod_{(\mathbf{f})} \binom{D_{\mathbf{f}}}{y_{\mathbf{f}}} \tag{11.63}$$

with $0 \leqslant y_{\mathbf{f}} \leqslant D_{\mathbf{f}}$; $\sum_{(\mathbf{f})} y_{\mathbf{f}} = n$. Moments and product-moments can be evaluated from (11.60) and (11.61) by the methods used in Chapter 2.

Variance of the estimators

The covariance matrix of $\tilde{\mathbf{D}}$ is

$$\mathbf{Var}(\tilde{\mathbf{D}}) = (N^{-1}n)^2\,\mathbf{P}^{-1}\,\mathrm{Var}(\mathbf{Z})\,(\mathbf{P}^{-1})^{\mathrm{T}} \tag{11.64}$$

(where the superscript T means transpose of a matrix). From Chapter 2,

$$\mathrm{var}(Z_{\mathbf{g}}) = n\,p_{\mathbf{g}} - n\{(n-1)\overline{p_{\mathbf{g}}^2} + (N-n)\bar{p}_{\mathbf{g}}^2\}\,(N-1)^{-1}; \tag{11.65a}$$

$$\mathrm{cov}(Z_{\mathbf{g}}, Z_{\mathbf{g}^*}) = -n\{(n-1)\,\overline{p_{\mathbf{g}}p_{\mathbf{g}^*}} + (N-n)p_{\mathbf{g}}p_{\mathbf{g}^*}\}\,(N-1)^{-1}, \tag{11.65b}$$

where

$$p_{\mathbf{g}} = N^{-1}\sum_{(\mathbf{f})} D_{\mathbf{f}}\,p_{\mathbf{g}|\mathbf{f}};$$

$$\overline{p_{\mathbf{g}}^2} = N^{-1}\sum_{(\mathbf{f})} D_{\mathbf{f}}\,p_{\mathbf{g}|\mathbf{f}}^2;$$

and

$$\overline{p_{\mathbf{g}}p_{\mathbf{g}^*}} = N^{-1}\sum_{(\mathbf{f})} D_{\mathbf{f}}\,p_{\mathbf{g}|\mathbf{f}}\,p_{\mathbf{g}^*|\mathbf{f}}.$$

Uniform accuracy case

Here let $p_j = p$ and $p'_j = p'$ for all j, that is the uniform inspection accuracy case. Let s_{ab}, where $a, b = 0$ or 1, be the number of type (j) nonconformity such that $g_j = a$ and $f_j = b$. Then we have

$$p_{\mathbf{g}|\mathbf{f}} = p^{s_{11}}p'^{s_{10}}(1-p)^{s_{01}}(1-p')^{s_{00}}.$$

For $k = 2$, we have

$$\mathbf{P} = \begin{pmatrix} (1-p')^2 & (1-p)(1-p') & (1-p)(1-p') & (1-p)^2 \\ p'(1-p') & p(1-p') & (1-p)p' & p(1-p) \\ p'(1-p') & (1-p)p' & p(1-p') & p(1-p) \\ p'^2 & pp' & pp' & p^2 \end{pmatrix}, \tag{11.66a}$$

where the order of rows and columns correspond to (gf) equal to 00, 01, 10 and 11, respectively. Therefore

$$(p-p')^2\mathbf{P}^{-1} = \begin{pmatrix} p^2 & -p(1-p) & -p(1-p) & (1-p)^2 \\ -pp' & p(1-p') & (1-p)p' & -(1-p)(1-p') \\ -pp' & (1-p)p' & p(1-p') & -(1-p)(1-p') \\ p'^2 & -p'(1-p') & -p'(1-p') & (1-p')^2 \end{pmatrix}. \tag{11.66b}$$

Hence

$$\begin{aligned}
\tilde{D}_{00} &= (N n^{-1})(p-p')^{-2}\{p^2 Z_{00} - p(1-p)(Z_{01}+Z_{10}) \\
&\quad + (1-p)^2 Z_{11}\}; \\
\tilde{D}_{01} &= (N n^{-1})(p-p')^{-2}\{-pp'Z_{00} + p(1-p')Z_{01} \\
&\quad + (1-p)p'Z_{10} - (1-p)(1-p')Z_{11}\}; \\
\tilde{D}_{10} &= (N n^{-1})(p-p')^{-2}\{-pp'Z_{00} - (1-p)p'Z_{01} \\
&\quad + p(1-p')Z_{10} - (1-p)(1-p')Z_{11}\}; \\
\tilde{D}_{11} &= (N n^{-1})(p-p')^{-2}\{p'^2 Z_{00} - p'(1-p')(Z_{01}+Z_{10}) \\
&\quad + (1-p')^2 Z_{11}\}.
\end{aligned} \tag{11.67}$$

Note that there is a formula for the calculation of $\mathbf{P}^{-1}$. For $k=2$,

$$\mathbf{P} = \mathbf{D} \otimes \mathbf{D}$$

with

$$\mathbf{D} = \begin{pmatrix} 1-p' & 1-p \\ p' & p \end{pmatrix}.$$

where $\otimes$ denotes the direct (Kronecker) product. Therefore

$$\mathbf{P}^{-1} = \mathbf{D}^{-1} \otimes \mathbf{D}^{-1},$$

with

$$\mathbf{D}^{-1} = (p-p)^{-1}\begin{pmatrix} p & -(1-p \\ -p' & 1-p' \end{pmatrix}.$$

Similarly for general k

$$\mathbf{P}^{-1} = \mathbf{D}^{-1} \otimes \mathbf{D}^{-1} \otimes \cdots \otimes \mathbf{D}^{-1}, \quad (k \text{ factors})$$

with the rows and columns rearranged in binary order.

We could obtain results for $k=3$ in a similar manner. For example

$$\begin{aligned}
\tilde{D}_{011} &= (N n^{-1})(p-p')^{-3}\{pp'^2 Z_{000} - pp'(1-p')(Z_{001}+Z_{010}) \\
&\quad - (1-p)p'^2 Z_{100} + p(1-p')^2 Z_{011} \\
&\quad + (1-p)p'(1-p')(Z_{101}+Z_{110}) - (1-p)(1-p')^2 Z_{111}\}.
\end{aligned} \tag{11.68}$$

Further special cases

Suppose in addition, we also assume that $D_{\mathbf{f}} = 2^{-k}N$ for all $(\mathbf{f})$. Let $m(\equiv (\mathbf{g}))$ be the number of 1's in $(\mathbf{g})$ and $m_{ab}(\equiv m_{ab}(\mathbf{g}, \mathbf{g}^*))$ be the number of NC type (**j**) with $g_j = a$ and $g_j^* = b$. ($m_{11} + m_{10} + m_{01} + m_{00} = k$.) We then have

$$\bar{p}_{\mathbf{g}} = 2^{-k}(p + p')^m (2 - p - p')^{k-m}, \tag{11.69a}$$

$$\overline{p_{\mathbf{g}}^2} = 2^{-k}(p^2 + p'^2)^m \{(1-p)^2 + (1-p')^2\}^{k-m} \tag{11.69b}$$

and

$$\overline{p_{\mathbf{g}} p_{\mathbf{g}^*}} = 2^{-k}(p^2 + p'^2)^{m_{11}} \{(p(1-p') + (1-p)p'\}^{m_{01}+m_{10}} \times \{(1-p)^2 + (1-p')^2\}^{m_{00}}. \tag{11.69c}$$

We further assume that $p' = (1-p)$; that means error probabilities are the same for any type of incorrect classification. Similarly

$$p_{\mathbf{g}} = 2^{-k}, \tag{11.70a}$$

$$\overline{p_{\mathbf{g}}^2} = 2^{-k}(p^2 + (1-p)^2)^k \tag{11.70b}$$

and

$$\overline{p_{\mathbf{g}} p_{\mathbf{g}^*}} = 2^{-k}(p^2 + (1-p)^2)^{m_{00}+m_{11}} \{2p(1-p)\}^{m_{01}+m_{10}}. \tag{11.70c}$$

Let

$$\psi \equiv m_{00} + m_{11}$$

be the number of j's such that $g_j = g_j^*$, and

$$A \equiv p^2 + (1-p)^2; \qquad B \equiv 1 - A = 2p(1-p).$$

By (11.65a,b) and (11.69a–c)

$$\operatorname{var}(Z_{\mathbf{g}}) = n 2^{-k} \left(1 - \frac{N-n}{N-1} 2^{-k} - \frac{n-1}{N-1} A^k \right); \tag{11.71a}$$

$$\operatorname{cov}(Z_{\mathbf{g}}, Z_{\mathbf{g}^*}) = -n 2^{-k} \left(\frac{N-n}{N-1} 2^{-k} + \frac{n-1}{N-1} A^{\psi} B^{k-\psi} \right). \tag{11.71b}$$

For $k = 3$ case, under the above special assumptions, we obtain

$$\tilde{D}_{001} = (N n^{-1})(2p-1)^{-6} \{ p(1-p)^2 (Z_{000} + Z_{101} + Z_{110}) - p^2(1-p)(Z_{111} + Z_{001} + Z_{010}) + p^3 Z_{011} - (1-p)^3 Z_{100} \}; \tag{11.27a}$$

$$\operatorname{var}(\tilde{D}_{001}) \doteq \frac{1}{n} \frac{N^2}{8} \left\{ \frac{A^3}{(2p-1)^6} - \frac{N - 9n + 8}{8(N-1)} \right\}. \tag{11.72b}$$

11.6 Testing hypothesis $p = 1$ (assuming $p' = 0$)

Suppose m successive independent random samples of the same size n are taken (and each sample is returned to the lot after inspection) from a lot of size N containing D NC items (we will use symbols $f \equiv n/N$ and $\omega \equiv D/N$). These samples will be inspected with the same probability p of correct indication of an NC item. We assume there can be no false indication of nonconformity, i.e. $p' = 0$. Let $Z_1, \ldots, Z_m$ be the numbers of items declared NC from the m samples, respectively. Our purpose is to test the null hypothesis: $p = 1$. Note that if $Z_1, \ldots, Z_m$ were from m different lots with the same proportion of NC items, the following analysis would also be applicable.

Distribution

We know that each of the Z's has the hypergeometric–binomial distribution (using Z as a generic term for any one of the Z's).

$$\Pr[Z = z|D, p] = \binom{N}{n}^{-1} \sum_{y \geq z} \binom{D}{y}\binom{N-D}{n-y}\binom{y}{z} p^z(1-p)^{y-z} \tag{11.73}$$

with $\max(0, n-N+D) \leqslant z \leqslant \min(n, D)$. The mean ($\mu$) and variance ($\sigma^2$) of Z are

$$\mu = np\frac{D}{N} (= np\omega = N\,pf\omega) \tag{11.74a}$$

and

$$\begin{aligned}\sigma^2 &= n\left\{\frac{pD}{N}\left(1 - \frac{pD}{N}\right) - \frac{n-1}{N-1}\frac{D}{N}\left(1 - \frac{D}{N}\right)p^2\right\} \\ &= n\left\{p\theta(1 - p\omega) - \frac{n-1}{N-1}p^2\omega(1-\omega)\right\} \\ &= \frac{N-n}{N-1}\mu(1 - n^{-1}\mu) + \frac{n-1}{N-1}(1-p)\mu \\ &\doteq (1-f)\mu(1-n^{-1}\mu) + f(1-p)\mu\end{aligned} \tag{11.74b}$$

The r-th descending factorial moment is

$$\mu_{(r)} = E[Z^{(r)}] = E[Z(Z-1)\cdots(Z-r+1)] = n^{(r)}D^{(r)}p^r/N^{(r)}. \tag{11.74c}$$

The testing problem

The null hypothesis is $H_0{:}p = 1$ with D as a nuisance parameter. We want to find a function of σ^2 and μ, that depends only on p. Then we can construct a test statistic based on that function. The following approach is based on Johnson and Kotz (1985a). From (11.74a, b), the function

$$\Omega \equiv \sigma^2\mu^{-1} + n^{-1}(N-n)\mu$$
$$= 1 - (N-1)^{-1}(n-1)p \tag{11.75}$$

is appropriate. With μ and σ^2 replaced by their unbiased estimators $\bar{Z} = m^{-1}\sum_{i=1}^m Z_i$ and $S^2 = (m-1)^{-1}\sum_{i=1}^m (Z_i - \bar{Z})^2$, respectively, we have a test statistic

$$W \equiv S^2 Z^{-1} + n^{-1}(N-1)^{-1}(N-n)Z. \tag{11.76}$$

The fact that

$$\Omega \begin{cases} = (N-1)^{-1}(N-n) & \text{if } p = 1\,(H_0), \\ > (N-1)^{-1}(N-n) & \text{if } p < 1\,(H_1), \end{cases}$$

suggests that when W is too large we should reject H_0. Of course, we have to find critical values, and the statistic W has to have a meaning (e.g. we must not have situations such as $Z = 0$ or $S = 0$). To construct a normal approximation, we will find expressions (approximations) for $E[W|D,p]$ and $\operatorname{Var}(W|D,p)$. Let

$$\sqrt{\beta_1} = \sigma^{-3}E[(Z-\mu)^3|D,p]; \qquad \beta_2 = \sigma^{-4}E[(Z-\mu)^4|D,p],$$

be the moment ratios of the distribution (11.73). The expected value is

$$E[W|D,p] = E[S^2\bar{Z}^{-1}|D,p] + N^{-1}(N-1)^{-1}(N-n)pD$$
$$\doteq \frac{\sigma^2}{\mu}\left\{1 - \frac{\operatorname{cov}(S^2,\bar{Z}|D,p)}{\sigma^2\mu} + \frac{\operatorname{Var}(\bar{Z}|-D,p)}{\mu^2}\right\}$$
$$+ n^{-1}(N-1)^{-1}(N-n)\mu$$
$$\doteq \frac{\sigma^2}{\mu}\left[1 - m^{-1}\left\{\sqrt{\beta_1}\frac{\sigma}{\mu} - \left(\frac{\sigma}{\mu}\right)^2\right\}\right]$$
$$+ n^{-1}(N-1)^{-1}(N-n)\mu$$
$$= \Omega - m^{-1}\sigma\left(\frac{\sigma}{\mu}\right)^2\left(\beta_1 - \frac{\sigma}{\mu}\right), \tag{11.77a}$$

and the variance is

$$\begin{aligned}\operatorname{var}(W|D,p) &= \operatorname{var}(S^2\bar{Z}^{-1}|D,p) + 2n^{-1}(N-1)^{-1}(N-n)\\ &\quad \times \operatorname{cov}(S^2\bar{Z}^{-1}, Z|D,p) + n^{-2}(N-1)^2 \operatorname{var}(\bar{Z}|D,p)\\ &\doteq \frac{\sigma^4}{\mu^2}\left\{\frac{\operatorname{var}(S^2|D,p)}{\sigma^2} - \frac{2\operatorname{cov}(S^2,\bar{Z}|D,p)}{\sigma^2\mu} + \frac{\operatorname{var}(\bar{Z}|D,p)}{\mu^2}\right\}\\ &\quad + 2n^{-1}(N-1)^{-1}(N-n)\left[\sigma^2 - \frac{\sigma^2}{\mu}\mu\left\{1 - m^{-1}\mu\right.\right.\\ &\quad \left.\left.\times\left(\sqrt{\beta_1} - \frac{\sigma}{\mu}\right)\right\}\right] + m^{-1}n^{-2}(N-1)^{-2}(N-n)^2\sigma^2\\ &\doteq \frac{\sigma^4}{m\mu^2}\left[\beta_2 - \frac{m-3}{m-1} - 2\frac{\sigma}{\mu}\sqrt{\beta_1} + \left(\frac{\sigma}{\mu}\right)^2 + 2n^{-1}(N-1)^{-1}\right.\\ &\quad \left.\times(N-n)\left(\frac{\mu}{\sigma}\sqrt{\beta_1} - 1\right) + n^{-2}(N-1)^{-2}(N-n)^2\left(\frac{\mu}{\sigma}\right)^2\right].\end{aligned} \tag{11.77b}$$

When H_0 is true, ($p=1$), Z is hypergeometric with parameters $(n; D, N)$, and Johnson and Kotz (1969, p. 144) give formulas for calculating $\sqrt{\beta_1}$ and β_2 in this case. For large N, we could use the approximation

$$(N-an)/(N+b) \doteq 1 - af$$

for fixed a and b. Therefore, with $n' \equiv n(1-f)$,

$$\sqrt{\beta_1} \doteq \frac{(1-2f)(1-2\omega)}{\sqrt{n'\omega(1-\omega)}};$$

$$\beta_2 \doteq 3 + \frac{1}{n'Q(1-\omega)}[1 - 6\{f(1-f) + \omega(1-\omega)\}]. \tag{11.78}$$

Furthermore

$$\mu = n\theta; \qquad \sigma^2 \doteq n'\omega(1-\omega). \tag{11.79}$$

Consequently,

$$E[W|D,1] \doteq \frac{N-n}{N-1} + \frac{(1-f)(1-\omega)}{mn'\omega}(f + \omega - 3f\omega); \tag{11.80a}$$

$$\operatorname{var}(W|D,1) \doteq \frac{(1-f)^2(1-\omega)^2}{m}\left[\frac{2m}{m-1}+\frac{1}{n'\omega(1-\omega)}\{1-6f(1-f)\right.$$
$$\left.-6\omega(1-\omega)-(1-f)(1-3f)(1-2\omega)^2\}\right]$$
$$=\frac{2(1-f)^2(1-\omega)^2}{m-1}+\frac{(1-f)(1-\omega)}{nm\theta}g(f,\omega) \quad (11.80\text{b})$$

with $g(f,\omega)=3f^2-2f+2\omega(1-\omega)(6f^2-8f-1)$. Since $6f^2-8f-1<0$ for $0\leqslant f\leqslant 1$, and noticing $0\leqslant\omega\leqslant 1$, we have, for fixed f,

$$-\tfrac{1}{2}-6f(1-f)\leqslant g(f,\omega)\leqslant -2f+3f^2, \quad (11.81)$$

and therefore

$$-\tfrac{1}{2}\leqslant g(f,\omega)\leqslant 1.$$

Taking the first terms of (11.80a) and (11.80b), with nuisance parameter $\omega=0$, we have a conservative test (based on normal approximation) with a critical region

$$W>(N-1)^{-1}(N-n)+z_{1-\alpha}(1-f)\left(\frac{2}{m-1}\right)^{1/2} \quad (11.82\text{a})$$

or

$$W>(1-f)\left[1+z_{1-\alpha}\left(\frac{2}{m-1}\right)^{1/2}\right] \quad (11.82\text{b})$$

where $z_{1-\alpha}$ is the upper $100\alpha\%$ point of the standard normal distribution.

For obtaining a rough power function (with respect to p), we take W to be normally distributed with mean $(1-fp)$ and standard deviation $(1-fp)\sqrt{2/(m-1)}$. The approximate power function is then

$$1-\Phi\left(\frac{1-f}{1-fp}z_{1-\alpha}-\frac{f(1-p)}{1-fp}\sqrt{\frac{m-1}{2}}\right), \quad (11.83)$$

where Φ is the standard normal cumulative distribution function. In order to get a (approximate) power greater than $100\beta\%$, we must have

$$\frac{1-f}{1-fp}z_{1-\alpha}-\frac{f(1-p)}{1-fp}\sqrt{\frac{m-1}{2}}<-z_\beta,$$

or

$$m > 2f^{-2}(1-p)^{-2}\{(1-f)z_{1-\alpha} + (1-fp)z_\beta\}^2 + 1. \qquad (11.84)$$

If $\alpha = 0{\cdot}1$ and $\beta = 0{\cdot}9$, we would need

$$m > 3{\cdot}2830 f^{-2}(1-p)^{-2}\{2 - f(1+p)\}^2 + 1,$$

giving, for example,

$$m > \begin{cases} 82{\cdot}075 f^{-2}(2 - 1{\cdot}8f)^2 + 1 & \text{for } p = 0{\cdot}8 \\ 36{\cdot}478 f^{-2}(2 - 1{\cdot}7f)^2 + 1 & \text{for } p = 0{\cdot}7. \end{cases}$$

The test is conservative because setting $\omega = 0$ implies the greatest possible variance of W. The required m might be impractically large unless f is quite large. (Because ω is usually unknown, it is more or less true to say that we are attempting to draw a distinction between a binomial and a hypergeometric distribution.) Since the differences are slight if the lot size is not small, it is to be expected that very large – often prohibitively so – sample sizes are needed to detect them.

Remarks on other approaches

It is easy to construct a test of H_0: $p = 1$, when we are inspecting items known to be NC. We could reject H_0 as soon as an NC item is indicated. This test would have level zero and power function $(1 - p^m)$, m being the number of items to be tested. Of course this test is not applicable if the inspection causes destruction of an NC item or if no NC items exist in the lot.

Sometimes, when n is quite large (and therefore it is quite likely that there are real NC items in the sample) and it is not feasible to trace individual items, we may inspect the same sample m times and record the total numbers of declared NC items, $(T_1, \ldots, T_m)$, say. Any difference between any two of the T's indicates $p < 1$. In fact, for sampling without replacement from a lot of size N which has D NC items, given that there exist y NC items in the sample, the conditional probability there are at least two different T's is

$$1 - \sum_{t=0}^{y} \left\{ \binom{y}{t} p^t (1-p)^{y-t} \right\}^m.$$

Taking expectation with respect to Y we obtain the overall probability

$$1 - \Pr[T_1 = T_2 = \cdots = T_m] = 1 - E_Y\left[\sum_{t=0}^{Y}\left\{\binom{Y}{t}p^t(1-p)^{Y-t}\right\}^m\right]$$

where

$$Y \sim \mathrm{Hypg}(n, D, N).$$

CHAPTER 12

Stratified populations: grading

We have already encountered stratified populations in Chapter 2 (section 2.4). The strata were differentiated according to probabilities of correct identification of C and NC items with respect to a single type of nonconformity. In other circumstances, stratification can refer to multitype nonconformities, as in Chapter 5. More generally, we can consider problems arising in the assignment of items to appropriate strata, whether or not these are defined in terms of nonconformities.

In this chapter, we consider two types of problem arising with stratified populations: grading and quota fulfilment.

12.1 Grading

Grading of products is one of the most basic human activities, predating even counting and ranking. It is still one of the basic tools of quality assessment – among many examples we note grading of municipal bonds according to solvency, and of eggs according (broadly) to size.

Problems arising from errors in grading have received relatively little attention in quality assurance literature. The theory presented in this section and the following one is based almost exclusively on research in which the authors of this book were involved (Johnson and Kotz (1985a), Kotz and Johnson (1983), briefly summarized in the review paper Johnson *et al.* (1990), where additional tables are presented).

Suppose there is a lot of size N, which contains s different classes (strata), $\Pi_1,\ldots,\Pi_s$ with $N_1,\ldots,N_s$ items, respectively, ($\sum_{i=1}^{s} N_i = N$, of course). Our aim is to assign an item to one of these s classes. Let P_{ij} be the probability that a real Π_j item will be classified as Π_i. Note that $\sum_{i=1}^{s} P_{ij} = 1$.

Suppose a random sample of size n is taken from the lot, and let

Y_j be the number of items belonging to Π_j in the sample ($\sum_{j=1}^{s} Y_j = n$). Let Z_{ij} denote the number, among these Y_j, assigned to Π_i ($i = 1,\ldots,s$). Using results in Chapter 3 $\mathbf{Y}(\equiv(Y_1,\ldots,Y_s))$ has a multivariate hypergeometric distribution with parameters $(n;\mathbf{N};N)$, so that

$$\Pr(\mathbf{Y}=\mathbf{y}) = \left\{\prod_{j=1}^{s}\binom{N_j}{y_j}\right\}\Big/\binom{N}{n} \qquad \text{where}\left(\sum_{j=1}^{s} y_j = n\right), \tag{12.1}$$

or, symbolically,

$$\mathbf{Y} \sim \text{Mult Hypg}(n;\mathbf{N};N),$$

where $\mathbf{N} \equiv (N_1,\ldots,N_s)$. Also, given $\mathbf{Y}$, a vector $\mathbf{Z}_j = (Z_{1j},\ldots,Z_{sj})$ has a multinomial distribution with parameters $(Y_j, \mathbf{P}_j)$ where $\mathbf{P}_j \equiv (P_{1j},\ldots,P_{sj})$, and

$$\Pr(\mathbf{Z}_j = \mathbf{z}_j | \mathbf{Y} = \mathbf{y}) = y_j! \prod_{i=1}^{s}\left(\frac{P^{z_{ij}}}{z_{ij}!}\right) \qquad \text{where} \sum_{i=1}^{s} z_{ij} = y_j. \tag{12.2}$$

Also $\mathbf{Z}_1,\ldots,\mathbf{Z}_s$ are mutually independent. The joint distribution of $\mathbf{Z}(\equiv(\mathbf{Z}_1,\ldots,\mathbf{Z}_s))$ is

$$\Pr(\mathbf{Z}=\mathbf{z}) = \sum_{\mathbf{y}} \Pr(\mathbf{Z}=\mathbf{z}|\mathbf{Y}=\mathbf{y})\Pr(\mathbf{Y}=\mathbf{y})$$

$$= \binom{N}{n}^{-1} \prod_{j=1}^{s}\left\{N_j^{(z\cdot j)} \prod_{i=1}^{s}\left(\frac{P^{z_{ij}}}{z_{ij}}\right)\right\}, \tag{12.3}$$

where $z_{.j} = \sum_{i=1}^{s} z_{ij}$. Note that the Y_j's are determined by the z_{ij}'s.

The joint distribution of $Z_{1.},\ldots,Z_{s.}$, where $Z_{i.} = \sum_{j=1}^{s} Z_{ij}$, is

$$*\cdots*_{j=1}^{s} \text{ Multinom}(Y_j, \mathbf{P}_j) \bigwedge_{\mathbf{Y}} \text{Mult Hypg}(n;\mathbf{N},N), \tag{12.4}$$

a so-called 'multivariate hypergeometric–multinomial' distribution.

The joint factorial moments of $\mathbf{Z}$ can be obtained from

$$E\left(\prod_{i=1}^{s}\prod_{j=1}^{s} Z_{ij}^{(r_{ij})}\right) = E_{\mathbf{Y}}\left|E\left(\prod_{j=1}^{s}\prod_{i=1}^{s} Z_{ij}^{(r_{ij})}|\mathbf{Y}\right)\right|$$

$$= E\left(\prod_{j=1}^{s} Y_j^{(r_{\cdot j})} \prod_{i=1}^{s} P_{ij}^{r_{ij}}\right)$$

$$= \frac{n^{(r_{..})}}{N^{(r_{..})}} \prod_{j=1}^{s}\left\{N_j^{(r_{.j})} \prod_{i=1}^{s} P_{ij}^{r_{ij}}\right\}, \tag{12.5}$$

where $r_{.j}=\sum_{i=1}^{s} r_{ij}, r_{..}=\sum_{i=1}^{s}\sum_{j=1}^{s} r_{ij}$. For $i \neq i', j \neq j'$, we have

$$E[Z_{ij}] = nN^{-1}N_jP_{ij}; \tag{12.6a}$$

$$E[Z_{ij}Z_{i'j}] = n^{(2)}N_j^{(2)}P_{ij}P_{i'j}/N^{(2)}; \tag{12.6b}$$

$$E[Z_{ij}Z_{i'j'}] = n^{(2)}N_jN_{j'}P_{ij}P_{i'j'}/N^{(2)}, \tag{12.6c}$$

from which

$$\begin{aligned}\operatorname{cov}(Z_{ij}, Z_{i'j}) &= -nN_j\{(N-n)N_j + N(n-1)\}P_{ij}P_{i'j}N^{-2}(N-1)^{-1};\\ \operatorname{cov}(Z_{ij}, Z_{i'j'}) &= -nN_jN_{j'}(N-n)P_{ij}P_{i'j'}N^{-2}(N-1)^{-1}.\end{aligned} \tag{12.7}$$

These covariances are negative, as might be expected, since there are constraints on sums of the Z's, so that the greater one Z is, the smaller the remainder tend to be.

Denote

$$\bar{P}_i = N^{-1}\sum_{j=1}^{s} N_jP_{ij}, \qquad (\overline{P_iP_{i'}}) = N^{-1}\sum_{j=1}^{s} N_jP_{ij}P_{i'j}.$$

Note the interpretations of $\bar{P}_i$ and $(\overline{P_iP_{i'}})$ as

$$\begin{aligned}\bar{P}_i &= \Pr[\text{a randomly chosen item is assigned to } \Pi_i];\\ (\overline{P_iP_{i'}}) &= \Pr[\text{a randomly chosen item is assigned to } \Pi_i \text{ and } \Pi_{i'}\\ &\qquad \text{in two independent tests}].\end{aligned}$$

We then have the covariance

$$\operatorname{cov}(Z_{i.}, Z_{i'.}) = -n\{(N-n)P_iP_{i'} + (n-1)(\overline{P_iP_{i'}})\}(N-1)^{-1}, \tag{12.8}$$

where, as before,

$$Z_{i.} = \sum_{j=1}^{s} Z_{ij} \qquad \text{and} \qquad Z_{i'.} = \sum_{j=1}^{s} Z_{i'j}$$

are the total number of items classified to $\Pi_i, \Pi_{i'}$ respectively.

12.2 A grading problem on multitype nonconformities

We now continue our investigation of NC patterns initiated in Chapter 5, with special reference to grading based on the number of types of nonconformity present in an item.

The assumptions here are similar to the previous section. Suppose there are s NC types. An item is graded by the total number of NC types which are ascribed to it. Let Z_g be the number of items in the sample judged to have NC pattern (**g**), where $\mathbf{g}=(g_1,\ldots,g_s)$ $g_i=0,1$ (cf. Chapter 5). For example, if $s=4$, Z_{1011} has pattern (1011), and is the number of items assigned to NC types 1, 3 and 4. But these Z_{1011} items should be graded according to the total number, $3=1+0+1+1$, of NC types they were assigned. We call such a grade C_3, or generally C_i, with $s\geqslant i$. The number of items assigned to C_j is

$$\sum\cdots\sum_{[j]} Z_{\mathbf{g}},$$

where $\sum\cdots\sum_{[j]}$ denotes summation over all **g** with $\sum_{i=1}^{s} g_i=j$. For instance, when $s=4$, the number of items assigned to C_3 is

$$Z_{0111}+Z_{1011}+Z_{1101}+Z_{1110}.$$

If $p_i=p$ and $p_i'=p'$ for all i, then our problem is the same as that studied in the previous section (12.1), with the following modifications:

1. The symbol N_j of section 12.1 is defined by
$$N_j=\sum\cdots\sum_{[j]} D_{\mathbf{g}}, \qquad \text{and} \tag{12.9}$$
2. the probability of assigning a C_j item to C_i is
$$P_{ij}=\Pr[V_1+V_0=i|C_j] \qquad (i,j=0,1,\ldots,s) \tag{12.10}$$
where V_1 and V_0 are mutually independent with
$$V_1|C_j\sim \text{Bin}(j,p);$$
$$V_0|C_j\sim \text{Bin}(s-j,p').$$

In a random sample of size n, let Z_{ij} denote the number of C_j items assigned to C_i, as in section 12.1. Directly from (12.4)–(12.8) (but i could be 0), we have the following result with j, $i=0,\ldots,s$:

$$\mathbf{Z}\sim(*\cdots*_{j=1}^{s}\ \text{Multinomial}(Y_j,\mathbf{P}_j))\ \underset{Y}{\wedge}\ \text{Multi Hypg}(n;\mathbf{N},N) \tag{12.11}$$

where $\mathbf{P}_j=(P_{0j},\ldots,P_{sj})$, $\mathbf{N}=(N_0,\ldots,N_s)$, and $\mathbf{Z}=(Z_{1\cdot},\ldots,Z_{s\cdot})$ with $Z_{i\cdot}=\sum_{j=0}^{s} Z_{ij}$. Note that $Z_{i\cdot}(Z_{i'\cdot})$ is the total number of items assigned to C_i, $(C_{i'})$. Also we have the covariance (cf. (5.49))

$$\text{cov}(Z_{i\cdot},Z_{i'\cdot})=-n\{(N-n)\bar{P}_i\bar{P}_{i'}+(n-1)\overline{(P_iP_{i'})}\}(N-1)^{-1} \tag{12.12}$$

with $\quad \bar{P}_i=N^{-1}\sum_{j=1}^{s} N_jP_{ij}, \qquad \overline{(P_iP_{i'})}=N^{-1}\sum_{j=1}^{s} N_jP_{ij}P_{i'j}.$

An interesting generalization of the problem discussed above arises if different weights (w_i) are assigned to nonconformities of different types (i) and grading is based on the sum of the weights. (If the weights are integers then the condition $\sum_{i=1}^{s} g_i = j$ (for assignment to C_j) would become $\sum_{i=1}^{s} w_i g_i = j$.)

12.3 Quota fulfilment with errors in stratification

The problem discussed in this section was investigated by Johnson (1957), assuming inspection to be perfect. The analysis incorporating an allowance for inspection errors was published by the authors of this book in Johnson *et al.* (1989). The presentation here follows the lines of that paper. A noteworthy feature is the introduction of cost functions, which we have ignored in most of the applications in this book, which emphasizes rather probabilities of correct classification. Reduction of the expected amount of inspection is, however, closely associated with cost.

Model and notation

Suppose that a population of effectively infinite size has k strata, $\Pi_1, \ldots, \Pi_k$, and the proportions of items in these strata are $p_1, \ldots, p_k$, respectively, with $p_1 + \cdots + p_k = 1$. Our aim is to take a random sample containing specified numbers $n_1, \ldots, n_k$ of items belonging to corresponding strata, respectively. The inspection is not perfect, and we have a cost structure. We will study how to minimize the expected cost. The sampling procedure is as follows.

1. We take a first random sample of size N from the lot and classify each item to the stratum it belongs by (imperfect) inspection.
2. For each $i = 1, \ldots, k$, if the number M_i, say, of items assigned (maybe with errors) to the ith stratum Π_i is less than n_i, then take further $(n_i - M_i)$ items (without error) from a set known to belong to Π_i to make good the shortfall.

Let P_{ij} have the same meaning as in section 12.1, i.e. the probability of assigning a true Π_j item to Π_i, and let Y_j be the actual number of true Π_j items in the first sample. Clearly

$$\sum_{i=1}^{k} M_i = \sum_{i=1}^{k} Y_i = N.$$

We assume that results of inspections of items are mutually independent with respect to strata, and

$$Y_i \sim \text{Bin}(N, p_i);$$
$$M_i \sim \text{Bin}(N, \omega_i),$$

where $\omega_i = \sum_{j=1}^{k} p_j P_{ij}$, the probability of a randomly chosen item being assigned to Π_i. Also conditionally

$$M_i | Y_i \sim \text{Bin}(Y_i, P_{ii}) * \text{Bin}(N - Y_i, \omega_i'),$$

where

$$\omega_i' = (1 - p_i)^{-1} \sum_{j \neq i}^{k} p_j P_{ij} = (1 - p_i)^{-1}(\omega_i - p_i P_{ii}).$$

is the probability that a randomly chosen item, not in Π_2, will be assigned to Π_i.

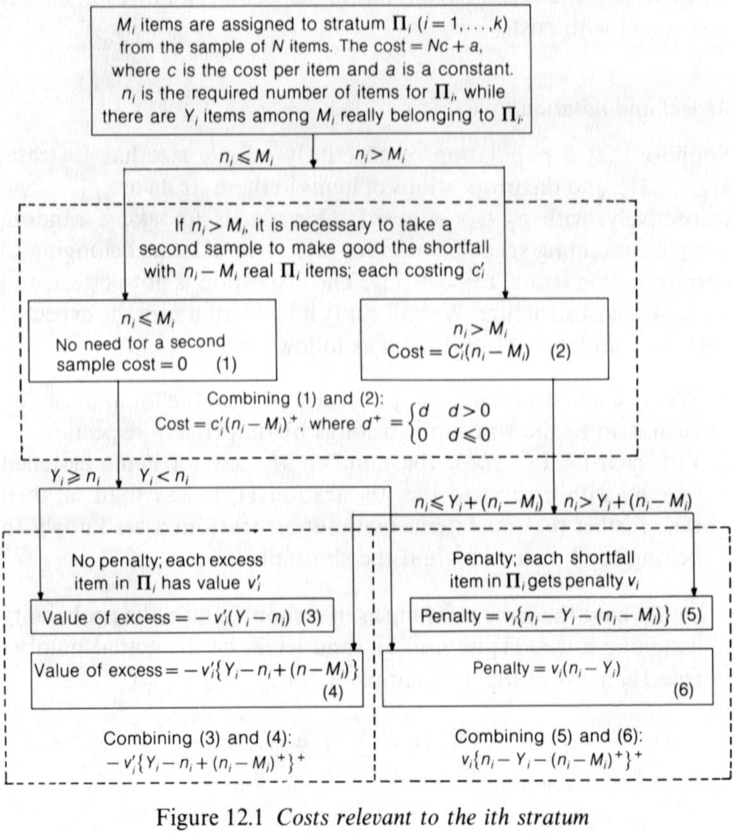

Figure 12.1 *Costs relevant to the ith stratum*

Costs

One cannot expect to have $M_i = Y_i$ for all i in step 2. The initial sample may be deficient or in excess in numbers for any given stratum. We introduce the following cost functions:

v_i: a 'penalty' for each item lacking from Π_i;

v'_i: the value (if any) of each item from Π_i in excess of the required number, n_i.

Let c be the sampling cost per item in the first step and c'_i the sampling cost per Π_i item in the second step, and let a be a constant cost added for the entire process. With the symbol

$$z^+ = \begin{cases} z & \text{if } z \geqslant 0 \\ 0 & \text{if } z < 0 \end{cases},$$

we have components of cost as set out below.

COMPONENT	COST
Obtaining first sample	$a + cN$
Obtaining second sample	$\sum_{i=1}^{k} c'_i(n_i - M_i)^+$
Penalty	$\sum_{i=1}^{k} v_i\{n_i - Y_i - (n_i - M_i)^+\}^+$
Value of excess	$-\sum_{i=1}^{k} v'_i\{Y_i - n_i + (n_i - M_i)^+\}^+$

The expected total cost is

$$\begin{aligned} C_N &= a + cN \\ &+ \sum_{i=1}^{k} \{v_i E[n_i - Y_i \mid Y_i < n_i] \Pr[Y_i < n_i] \\ &- v'_i E[Y_i - n_i \mid Y_i \geqslant n_i] \Pr[Y_i \geqslant n_i]\} \\ &+ \sum_{i=1}^{k} c_i E[n_i - M_i \mid M_i < n_i] \Pr[M_i < n_i] \\ &- \sum_{i=1}^{k} [\{v_i E[n_i - Y_i \mid M_i \leqslant Y_i \leqslant n_i] \\ &+ v'_i E[Y_i - M_i \mid M_i \leqslant Y_i \leqslant n_i]\} \Pr[M_i \leqslant Y_i \leqslant n_i] \\ &+ v_i E[n_i - M_i \mid Y_i < M_i < n_i] \Pr[Y_i < M_i < n_i] \\ &+ v'_i E[n_i - M_i \mid M_i < n_i \leqslant Y_i] \Pr[M_i < n_i \leqslant Y_i]]. \end{aligned} \quad (12.13)$$

This is the sum of:

1. the first line of (12.13): the expected cost of sampling in step 1, including the shortfall penalties and value of excess;

2. the second line of (12.13): the expected cost of sampling in step 2; and
3. the rest of (12.13): the expected value of savings due to the result of the second step making up shortfalls. This is illustrated in Fig. 12.1.

Instead of attempting the arduous, if not impossible, task of direct minimization of C_N with respect to N, we attack the problem by considering the difference in cost if the sample size increases by one,

$$\Delta C_N = C_{N+1} - C_N.$$

In addition to the sampling cost c, supposing that the added item belongs to Π_i and has been assigned to Π_j (with probability $p_i P_{ji}$), the difference has the following components:

1. the 'cost' due to the value of the extra item, which is

$$\begin{cases} -v_i & \text{if } Y_i < n_i, \\ -v_i' & \text{if } Y_i \geqslant n_i; \end{cases}$$

2. if $M_j < n_j$, in addition to the saving (less cost) $-c_j'$ from step 2 the cost is due to the lower value (greater cost) of the extra item

$$\begin{cases} -v_j & \text{if } Y_j < n_j, \\ -v_j' & \text{for } Y_j \geqslant n_j. \end{cases}$$

Hence

$$\begin{aligned}
\Delta C_N &= c - \sum_{i=1}^{k} p_i\{v_i \Pr[Y_i < n_i] + v_i' \Pr[Y_i \geqslant n_i]\} \\
&\quad - \sum_{i=1}^{k} p_i[P_{ji}\{c_j' \Pr[M_j < n_j] - v_j \Pr[(M_j < n_j) \cap (Y_j < n_j)] \\
&\quad - v_j' \Pr[(M_j < n_j) \cap (Y_i \geqslant n_j)]\}] \\
&= c - \sum_{i=1}^{k} p_i\{v_i' + (v_i - v_i') \Pr[Y_i < n_i]\} \\
&\quad - \sum_{i=1}^{k} p_i\left[\sum_{j=1}^{k} P_{ji}\{(c_j' - v_j') \Pr[M_j < n_j] \right. \\
&\quad \left. - (v_j - v_j') \Pr[(M_j < n_j) \cap (Y_j < n_j)]\}\right],
\end{aligned} \tag{12.14}$$

with

$$\Pr[Y_i < n_i] = \sum_{y=0}^{n_i-1} \binom{N}{y} p_i^y (1-p_i)^{N-y};$$

$$\Pr[M_i < n_i] = \sum_{m=0}^{n_i-1} \binom{N}{m} \omega_i^m (1-\omega_i)^{N-m}$$

and

$$\begin{aligned}\Pr[(M_i<n_i)\cap(Y_i<n_i)] &= \sum_{y=0}^{n_i-1} \Pr[(M_i<n_i)\cap(Y_i=y)] \\ &= \sum_{y=0}^{n_i-1} \binom{N}{y} p_i^y (1-p_i)^{N-y} \\ &\quad \times \sum_{u+v<n_i} \binom{y}{u} P_{ii}^u (1-P_{ii})^{y-u} \\ &\quad \times \binom{N-y}{v} \omega_i'^v (1-\omega_i')^{N-y-v}.\end{aligned}$$

Since $\Delta C_0 < 0$ and ΔC_N increases with N, under the conditions $c_i' > (v_i, c) > v_i'$ (for all i), the optimal value of N would be the least value for which $\Delta C_N \geqslant 0$. The solution depends on the values of c, c_i', v_i and v_i' only through the ratios

$$\{c - v_i'\} : \{c_i' - v_i'\} : \{v_i - v_i'\}. \qquad (i = 1, \ldots k)$$

Special cases

If $v_i = v$, $v_i' = v'$ and $c_i' = c'$ for all $i = 1, 2, \ldots, k$, then the solution of $\Delta C_N = 0$ depends only on the ratios

$$c - v' : c' - v' : v - v'.$$

We have

$$\begin{aligned}\Delta C_N = c - v' - (c' - v') \sum_{i=1}^{k} p_i \sum_{j=1}^{k} P_{ji} \Pr[M_j < n_j] \\ - (v - v')\left[\sum_{i=1}^{k} p_i \left\{\Pr[Y_i < n_i] \right.\right. \\ \left.\left. - \sum_{j=1}^{k} P_{ji} \Pr[(M_j < n_j)\cap(Y_j < n_j)]\right\}\right]. \qquad (12.15)\end{aligned}$$

We further suppose that incorrect assignments are proportional to strata sizes, so that

$$P_{ji} = \begin{cases} P & \text{if } i = j; \\ (1-p_i)^{-1} p_j(1-P) & \text{if } i \neq j \quad (i, j = 1, \ldots, k). \end{cases}$$

Then

$$\omega_i = p_i \left\{ P + (1-P) \sum_{j \neq i} p_j (1-p_j)^{-1} \right\} \qquad (12.16a)$$

and

$$\omega'_i = (1-p_i)^{-1}(1-P)p_i \sum_{j \neq i} p_i(1-p_j)^{-1}. \qquad (12.16b)$$

The correlation between Y_i and M_j is, in general,

$$p_i\left(P_{ji} - \sum_{h=1}^{k} p_h P_{jh}\right)\left[p_i(1-p_i)\left(\sum_{h=1}^{k} p_h P_{jh}\right)\left(1 - \sum_{h=1}^{k} p_h P_{jh}\right)\right]^{-1/2}. \qquad (12.17)$$

In the completely symmetric case, when, in addition to the above conditions, for all i

$$p_i = \frac{1}{k}; \quad n_i = \frac{N}{k} (\equiv n);$$

both M_i and Y_i have the same binomial distribution with parameters (N, k^{-1}) and the correlation between Y_i and M_i is $(k-1)^{-1}(kP-1)$. For $j \neq i$

$$\omega'_j = P_{ji} = (k-1)^{-1}(1-P). \qquad (12.18a)$$

Also, for all i

$$\Pr[Y_i < n] = \Pr[M_i < n] \equiv P(n) = k^{-N} \sum_{y=0}^{n-1} \binom{N}{y} (k-1)^{N-y}; \qquad (12.18b)$$

$$\Pr[(Y_i < n) \cap (M_i < n)] \equiv P^*(n)$$
$$= k^{-N} \sum_{y=0}^{n-1} \binom{N}{y} \sum_{u+v<n} \binom{y}{u}\binom{N-y}{v}$$
$$\times P^u(1-P)^{y-u+v}(k-2+P)^{N-y-v}; \qquad (12.18c)$$

and

$$\Delta C_N = c - v' - (c' - v')P(n) - (v - v')\{P(n) - P^*(n)\}. \qquad (12.18d)$$

Table 12.1 *Optimal Values of N Ratios are* $c-v':c'-v':v-v'$; $N=\min\{n:\Delta C_n>0\}$

k	n	$1:3:\frac{1}{2}$ $P=0{\cdot}85$	$0{\cdot}70$	$1:3:2$ $0{\cdot}85$	$0{\cdot}70$	$1:5:\frac{1}{2}$ $0{\cdot}85$	$0{:}70$	$1:5:2$ $0{:}85$	$0{:}70$
2	25	53	53	54	55	56	56	57	57
2	50	104	105	106	107	109	109	110	110
2	100	206	207	209	210	212	213	213	214
3	17	55	55	56	57	59	60	60	61
3	33	105	105	107	108	111	111	112	113
3	67	210	210	213	214	218	218	220	221
4	13	57	57	59	59	62	63	63	64
4	25	107	108	109	111	115	115	116	117
4	50	211	211	214	215	221	221	223	224
5	10	55	56	57	58	62	62	63	63
5	20	108	109	111	112	117	117	118	119
5	40	211	213	216	217	224	224	226	227
5	100	520	521	525	528	538	539	542	543
10	5	57	58	60	61	67	67	69	70
10	10	111	112	115	117	125	125	127	128
10	20	217	218	222	225	236	236	239	240
10	50	529	530	537	541	557	558	562	564

Since P appears only in $P^*(n)$, one can see from (12.18d) that if $v-v'$ is relatively small compared with $c-v'$ and $c'-v'$, then P has little influence on the optimal values of N in this symmetric case. Table 1 of Johnson *et al.* (1989), which overlaps with Table 12.1, clearly demonstrates this point. We therefore could assume $P=1$ (error-free), for calculating the optimal N here except for some extreme situations such as $P=0$. However, P has a substantial effect on the minimized value of expected cost.

APPENDIX A

Note on computer programs

The following SAS programs, written by Dr. Robert N. Rodriguez of SAS Institute Inc., SAS Campus Drive, Cary NC 27513-8000, USA, will be available in the SAS Sample Library provided with SAS/QC software (Release 6·07, expected in late 1991).

A directory of these programs is provided in the Sample Library program named INSPTERR.

Chapter 2

1. Distribution of Z (apparent number of defectives in single-lot sampling) under imperfect inspection when the lot size is finite.
2. Distribution of Z under imperfect inspection where N (lot size) approaches infinity.

Chapter 4

1. Acceptance probabilities for double sampling under inspection error, assuming finite lot size.
2. Acceptance probabilities for link sampling procedure.
3. Acceptance probabilities for partial link sampling.

Chapter 6

Simple Dorfman screening: PC(C), EPR
Two-stage Dorfman procedures: PC(C), PC(NC), EPR
Three-stage Dorfman procedures: PC(C), PC(NC), EPR

Chapter 7

Dorfman–Sterrett procedures: EPR, PC(NC), PC(C)

Chapter 8

Curtailed Dorfman testing: PC(NC), PC(C), EPR
Both finite and infinite lot sizes

Chapter 9

Graff–Roeloffs procedures: EPR

APPENDIX B

The detection of defective members of large populations*

BY ROBERT DORFMAN

Washington, D.C.

The inspection of the individual members of a large population is an expensive and tedious process. Often in testing the results of manufacture the work can be reduced greatly by examining only a sample of the population and rejecting the whole if the proportion of defectives in the sample is unduly large. In many inspections, however, the objective is to eliminate all the defective members of the population. This situation arises in manufacturing processes where the defect being tested for can result in disastrous failures. It also arises in certain inspections of human populations. Where the objective is to weed out individual defective units, a sample inspection will clearly not suffice. It will be shown in this paper that a different statistical approach can, under certain conditions, yield significant savings in effort and expense when a complete elimination of defective units is desired.

It should be noted at the outset that when large populations are being inspected the objective of eliminating all units with a particular defect can never be fully attained. Mechanical and chemical failures and, especially, man-failures make it inevitable that mistakes will occur when many units are being examined. Although the procedure described in this paper does not directly attack the problem of technical and psychological fallibility, it may contribute to its partial solution by reducing the tediousness of the work and by making more elaborate and more sensitive inspections economically feasible.

* *Ann. Math. Statist.*, **14**, 436–40. Reprinted with the authors permission (see Section 6.1 for comments on this paper.)

In the following discussion no attention will be paid to the possibility of technical failure or operators' error.

The method will be described by showing its application to a large-scale project on which the United States Public Health Service and the Selective Service System are now engaged. The object of the program is to weed out all syphilitic men called up for induction. Under this program each prospective inductee is subjected to a 'Wasserman-type' blood test. The test may be divided conveniently into two parts:

1. A sample of blood is drawn from the man,
2. The blood sample is subjected to a laboratory analysis which reveals the presence or absence of 'syphilitic antigen.' The presence of syphilitic antigen is a good indication of infection.

When this procedure is used, N chemical analyses are required in order to detect all infected members of a population of size N.

The germ of the proposed technique is revealed by the following possibility. Suppose that after the individual blood sera are drawn they are pooled in groups of, say, five and that the groups rather than the individual sera are subjected to chemical analysis. If none of the five sera contributing to the pool contains syphilitic antigen, the pool will not contain it either and will test negative. If, however, one or more of the sera contain syphilitic antigen, the pool will contain it also and the group test will reveal its presence.* The individuals making up the pool must then be retested to determine which of the members are infected. It is not necessary to draw a new blood sample for this purpose since sufficient blood for both the test and the retest can be taken at once. The chemical analyses require only small quantities of blood.

Two questions arise immediately:

1. Will the group technique require fewer chemical analyses than the individual technique and, if so, what is the extent of the saving; and
2. What is the most efficient size for the groups?

Both questions are answered by a study of the probability of obtaining an infected group. Let

$p =$ the prevalence rate per hundred, that is the probability that a random selection will yield an infected individual. Then

*Diagnostic tests for syphilis are extremely sensitive and will show positive results for even great dilutions of antigen.

$1-p=$ the probability of selecting at random an individual free from infection. And

$(1-p)^n=$ the probability of obtaining by random selection a group of n individuals all of whom are free from infection. Then

$p'=1-(1-p)^n=$ the probability of obtaining by random selection a group of n with at least one infected member.

Further

$N/n=$ the number of groups of size n in a population of size N, so

$p'N/n=$ the expected number of infected groups of n in a population of N with a prevalence rate of p.

The expected number of chemical analyses required by the grouping procedure would be

$$E(T)=N/n+n(N/n)p'$$

or the number of groups plus the number of individuals in groups which require retesting.* The ratio of the number of tests required by the group technique to the number required by the individual technique is a measure of its expected relative cost. It is given by:

$$C=T/N=1/n+p'$$

$$=\frac{n+1}{n}-(1-p)^n.$$

The extent of the saving attainable by use of the group method depends on the group size and the prevalence rate. Figure 1 shows the shape of the relative cost curve for five prevalence rates ranging from ·01 to ·15.† For a prevalence rate of ·01 it is clear from the chart that only 20% as many tests would be required by group tests with groups of 11 than by individual testing. The attainable savings decrease as the prevalence rate increases, and for a prevalence rate

---

*The variance of T is $\sigma_T^2=nNp'(1-p')=nN[(1-p)^n-(1-p)^{2n}]$. The coefficient of variation of T becomes small rapidly as N increases.

†The prevalence rate of syphilis among the first million selectees and volunteers was ·0185 for whites and ·2477 for other races. Geographically, the prevalence rate for whites ranged from ·0505 in Arizona to ·0051 in Wisconsin. See Parran, Thomas and Vonderlehr, R.A., *Plain Words about Venereal Disease*, Reynal, and Hitchcock, New York.

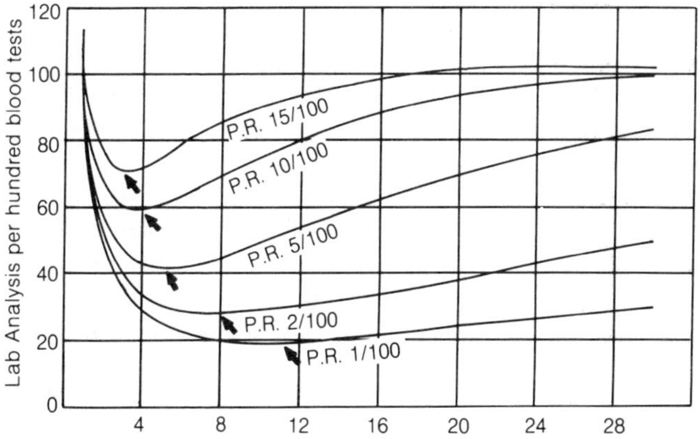

Figure 1 *Economies resulting from blood testings by groups. P.R. denotes prevalence rate*

Table I *Optimum group sizes and relative testing costs for selected prevalence rates.*

Prevalence rate (per cent)	Optimum group Size	Relative testing cost	Percent saving attainable
1	11	20	80
2	8	27	73
3	6	33	67
4	6	38	62
5	5	43	57
6	5	47	53
7	5	50	50
8	4	53	47
9	4	56	44
10	4	59	41
12	4	65	35
13	3	67	33
15	3	72	28
20	3	82	18
25	3	91	9
30	3	99	1

of ·15, 72% as many tests are required by the most efficient grouping as by individual testing. The optimum group size for a population with a known prevalence rate is the integral value of n which has the lowest corresponding value on the relative cost curve for that prevalence rate.

Optimum group sizes and their costs relative to the cost of individual testing are given in Table I for selected prevalence rates.

This table, together with the description of the group testing technique as it might be applied to blood tests for syphilis, reveals the two conditions for the economical application of the technique:

1. That the prevalence rate be sufficiently small to make worth while economies possible; and
2. That it be easier or more economical to obtain an observation on a group than on the individuals of the group separately.

Where these conditions exist, it will be more economical to locate defective members of a population by means of group testing than by means of individual testing.

The principle of group testing may be applied to situations where the interest centers in the degree to which an imperfection is present rather than merely in its presence or absence. For example, it could be applied to lots of chemicals where it is desired to reject all batches with more than a certain degree of impurity. If n samples of a chemical are pooled and subjected to a single analysis, the degree of impurity in the pool will be the average of the impurities in the separate samples. If the criterion were adopted that the members of a pool would be examined individually whenever the proportion of impurity in the pool is greater than $1/n$th the maximum acceptable degree of impurity, clearly no excessively impure batches would get by. The extent of the saving accomplished by this means can be computed by letting p' equal the probability that the pool will be impure enough to warrant retesting its constituent batches and using the formulas given above. The probability, p', can be calculated easily from the probability distribution of impurities in the separate batches.

It is evident that this approach will produce worthwhile savings only if the limit of acceptability is liberally above the per cent of impurity encountered in the bulk of the batches. It is also evident that under this scheme many of the retests will indicate that all the batches in the pool are acceptable and that the retesting was not really needed. The criterion for retesting can be raised above $1/n$th the limit of acceptability at the cost of a relatively small risk of accepting overly impure batches. The probability of failing to detect a defective batch when the retest criterion is raised in this manner will depend upon the form and parameters of the distribution of imperfection in single batches, as well as upon the number of batches in the pool. No simple general solution for this problem has been found.

References

Albers, W. and Veldman, H.J. (1984) Adaptive estimation of binomial probabilities under misclassification. *Statistica Neerlandica*, **38**, 223–247. (Ch. 11)

Armstrong, W. (1982) A chart to describe and measure the effects of individual inspection errors. *The Statistician, London*, **31**, 199–209. (Ch. 1)

Ayoub, M.M., Walvekar, A.G. and Lambert, B. K. (1970) *Statistical quality control under inspection errors.* Technical Paper IQ 70–284, Society of Manufacturing Engineers, Dearborn, Michigan. (Ch. 1)

Ballou, D.P. and Pazer, H.L. (1982) The impact of inspection fallibility on the inspection policy in serial production systems. *Mgmt. Sci.*, **28**, 387–399. (Ch. 1)

Bauer, L. (1987) *Inspektionsfehler in der Attributiven Qualitätskontrolle.* Physica Verlag, Heidelberg. (Ch. 1)

Bauer, L. (1990) Personal communication.

Beainy, I. and Case, K.E. (1981) A wide variety of AOQ and ATI performance measures with and without inspection error. *J. Qual. Technol.*, **13**, 1–9. (Ch. 1)

Bennett, G.K., Case, K.E. and Schmidt, J.W. (1974) The economic effects of inspector error on attribute sampling plans. *Naval Res. Logistic Qtly.*, **21**, 431–443. (Ch. 1)

Biegel, J.E. (1974) Inspection errors and sampling plans. *AIIE Trans.*, **6**, 284–287. (Ch. 1)

Bizley, M.T.L. (1962) Patterns in repeated trials, *J. Inst. Actu.*, **88**, 360–366. (Ch. 4)

Blischke, W.R. (1962) Moment estimators for the parameters of a mixture of two binomial distributions, *Ann. Math. Statist.*, **33**, 444–454. (Ch. 11)

Chapter numbers in parentheses refer to this volume.

Blischke, W.R. (1964) Estimating the parameters of mixtures of binomial distributions. *J. Amer. Statist. Ass.*, **59**, 510–528. (Ch. 11)

Bonett, D.G. (1988) Estimating the number of defects under imperfect inspection. *J. Appl. Statist.*, **15**, 63–67. (Ch. 5, 11)

Bross, I.D.J. (1954) Misclassification in 2 × 2 tables. *Biometrics*, **10**, 478–486. (Ch. 1)

Bureau of the Census, US (1964) *Evaluation and Research Program of the US Censuses of Population and Housing, 1960.* Series ER 60, Nos. 3, 4.(Ch. 1)

Carr, W.E. (1982) Sampling plan adjustment for inspection error and skip-lot plan. *J. Qual. Technol.*, **14**, 105–116. (Ch. 1)

Case, K.E., Bennett, G.K. and Schmidt, J.W. (1973) The Dodge CSP-1 continuous sampling plan under inspection error. *AIIE Trans.*, **5**, 193–202. (Ch. 1, 4)

Case, K.E., Bennett, G.K. and Schmidt, J.W. (1975) The effect of inspection error on average outgoing quality. *J. Qual. Technol.*, **7**, 28–33. (Ch. 1)

Chakraborty, S. and Bapaye, M.D. (1989) Effect of inspection error on MIL STD 105D sampling plans: some observations. *Intern. J. Qual. Rel. Mgmt.*, **6**, 60–70. (Ch. 4)

Chakraborty, S. and Mondal, A. (1985) Effect of inspection errors on choice of sampling plans. *Indian Ass. Prod. Qual. Rel. Trans.*, **10**, 29–43. (Ch. 1)

Chakraborty, S. and Rathie, U.K. (1989) Analysis of inspection error through GERT, *Sankhyā, Ser B*, **51**, 125–133. (Ch. 4)

Chen, C.L. and Swallow, W.H. (1990) Using group testing to estimate a proportion and to test the binomial model. *Biometrics*, **46**, 1035–1046.

Cleverley, D.S. (1983) Product quality level monitoring and control for logic chips of modules. *IBM J. Res. Dev.*, **27**, 4–10. (Ch. 1)

Cohen, A.C. (1960a) Misclassification in the Poisson distribution in which values of $(c + 1)$ are sometimes reported as c. *Ann. Inst. Statist. Math. (Tokyo)*, **11**, 189–193. (Ch. 1)

Cohen, A.C. (1960b) Misclassified data from a binomial population, *Technometrics*, **2**, 109–113. (Ch. 1)

Collins, R.D. (1974) *Statistically and economically based attribute acceptance sampling models with inspection errors.* PhD dissertation, Dept. Industrial Engineering and Operations Research, Virginia Polytechnic Institute and State University, Blacksburg Virginia. (Ch. 1)

Collins, R.D. and Case, K.E. (1976) The distribution of observed defectives in attribute acceptance sampling plans under inspection error. *AIIE Trans.*, **8**, 375–378. (Ch. 1)

Dahiya, R.C. (1981) An improved method of estimating an integer-parameter by maximum likelihood. *The American Statistician*, **35**, 34–37. (Ch. 11)

Das, M.N. and Choudhury, D.R. (1987) On problems of search using group tests, *Indian J. Statist.*, **49B**(2), 137–147. (Ch. 10)

Davis, A.L., Hoag, L.L. and Kasiviswanathan, G. (1977) Inspection error models, an empirical investigation. *Proc. 21st Annual Meeting, Human Factors Soc.*, pp. 137–141. (Ch. 1)

Davis, A.S. (1966) *The Sensitivity of Sampling Inspection to Inspector Error*, M.S. thesis, Texas Technological College, Lubbock, Texas. (Ch. 1)

Dempster, A.P., Laird, N.M. and Rubin, D.B. (1977) Maximum likelihood from incomplete data via the EM algorithm. *J. R. Statist. Soc. Ser. B.*, **39**, 1–22. (Ch. 11)

Diamond, E.L. and Lilienfeld, A.M. (1962) Misclassification errors in 2 × 2 tables with one margin fixed. *Amer. J. Publ. Hlth.*, **52**, 2106–2110. (Ch. 1)

Dodge, H.F. (1943) A sample inspection plan for continuous production, *Ann. Math. Statist.*, **14**, 264–279. (Ch. 4)

Dodge, H.F. (1970) Notes on the evolution of acceptance sampling plans IV. *J. Qual. Technol.*, **2**, 1–8. (Ch. 4)

Dodge, H.F. and Romig, H.G. (1929) A method of sampling inspection. *Bell System Tech. J.*, **8**, 613–631. (Ch. 1)

Dodge, H.F. and Torrey, M.N. (1951) Additional continuous sampling inspection plans. *Indust. Qual. Control*, **7**(5), 7–12. (Ch. 4)

Dorfman, R. (1943) The detection of defective members of large populations, *Ann. Math. Statist.*, **14**, 436–440. (Ch. 6)

Dorfman, R. (1991) Personal communication.

Dorris, A.L. and Foote, B.L. (1978) Effects of human inspector errors on statistical quality control plans. *Proc. AIIE Spring Conf.*, 487–492. (Ch. 1)

Drury, C.G. and Addison, J.L. (1973) An industrial study of the effects of feedback and fault density on inspection performance. *Ergonomics*, **16**, 159–169. (Ch. 1)

Garey, M.R. and Hwang, F.K. (1974) Isolating a single defective using group testing. *J. Amer. Statist. Ass.*, **67**, 151–153. (Ch. 10)

Gastwirth, J.L. and Hammick, P.A. (1989) Estimation of the

prevelance of a rare disease, preserving the anonymity of the subjects by group testing: Application to estimating the prevalence of AIDS antibodies in blood donors. *J. Statist. Plann. Inf.*, **22**, 15–27. (Ch. 6, 11)

Ghosh, D.T. (1985) An investigation on the effect of inspection error on AOQ of a single sampling acceptance verification system. *Sankhyā*, **47B**, 233–241. (Ch. 1)

Gill, A. and Gottlieb, D. (1974) The identification of a set by successive intersections. *Inf. Control*, **24**, 20–35. (Ch. 10)

Goldstein, I.L. and Mobley, W.H. (1971) Error and variability in the visual processing of critical radiographs. *J. Appl. Psychol.*, **55**, 549–553. (Ch. 1)

Graff, L.E. and Roeloffs, R. (1972) Group testing in the presence of test error: an extension of the Dorfman procedure. *Technometrics*, **14**, 113–122. (Ch. 9)

Graff, L.E. and Roeloffs, R. (1974) A group testing procedure in the presence of test error. *J. Amer. Statist. Assoc.*, **69**, 159–163. (Ch. 9)

Grassia, A. and Sundberg, R. (1982) Statistical precision in the calibration and use of sorting machines and other classifers. *Technometrics*, **24**, 117–121. (Ch. 1)

Gunther, W.C. (1969) Use of binomial, hypergeometric and Poisson tables to obtain sampling plans. *J. Qual. Technol.*, **4**, 105–109. (Ch. 1)

Guthrie, D. and Johns, M.V. (1959) Bayes acceptance sampling procedures for large lots. *Ann. Math. Statist.*, **30**, 896–925. (Ch. 1)

Hailey, W.A. (1980) Minimum sample size single sampling plans: a computerized approach, *J. Qual. Technol.*, **12**, 230–235. (Ch. 1)

Hammick, P.A. and Gastwirth, J.L. (1990) Estimation of the prevalence of AIDS and other sensitive characteristics by group testing in moderate prevalence populations. *Technical Report*, Dept. of Statistics, George Washington University, Washington, DC. (Ch. 6, 11)

Harishchandra, L. and Srivenkataramana, T. (1982) Link sampling for attributes. *Commun. Statist. – Theor. Meth.*, **11**, 1855–1868. (Ch. 4)

Harris, D.H. (1968) Effect of defect rate on inspection accuracy. *J. Appl. Psychol.*, **52**, 377–379. (Ch. 1)

Hayes, A.S. (1950) Control of visual inspection. *Indust. Qual. Control.*, **6**(6), 73–76. (Ch. 1)

Hoag, L.L., Foote, B.L. and Mount-Campbell, C. (1975) The effect of inspector accuracy on the Type I and Type II error of common sampling techniques. *J. Qual. Technol.*, **7**, 157–164. (Ch. 1)

Hochberg, Y. and Tenenbein, A. (1983) On triple sampling schemes for estimating from binomial data with misclassification errors. *Commun. Statist. – Theor. Meth.*, **12**, 1523–1533. (Ch. 11)

Huang, Q., Johnson, N.L. and Kotz, S. (1989) Modified Dorfman–Sterrett screening (group testing) procedures and the effects of faulty inspection. *Commun. Statist. – Theor. Meth.*, **18**, 1485–1495. (Ch. 7)

Hwang, F.K. (1974) On finding a single defective in binomial group testing. *J. Amer. Statist. Ass.*, **69**, 149–150. (Ch. 10)

Hwang, F.K. (1984) Robust group testing. *J. Qual. Technol.*, **16**, 189–195. (Ch. 1,6)

Jacobson, H.J. (1952) A study of inspector accuracy. *Indust. Qual. Control*, **9**, 16–25. (Ch. 1)

Johnson, N.L. (1957) Optimal sampling for quota fulfilment. *Biometrika*, **44**, 518–523. (Ch. 12)

Johnson, N.L. and Kotz, S. (1969) *Distributions in Statistics: Discrete Distributions*, Wiley, New York. (Ch. 8, 11)

Johnson, N.L. and Kotz, S. (1983) Faulty inspection distributions – some generalizations. In *Reliability in the Acquisitions Process* (eds. D.J. de Priest and R.L. Launer) Dekker, New York, (pp. 171–182). (Ch. 2, 3)

Johnson, N.L. and Kotz, S. (1984a) An estimation problem in faulty inspection sampling. *Coll. Math. Soc. Janos Bolyai*, **45**, 265–270. (Ch. 11)

Johnson, N.L. and Kotz, S. (1948b) Moment estimators of defect pattern frequencies, allowing for inspection error. *Commun. Statist.–Ther. Meth.*, **13**, 1215–1226. (Ch. 11)

Johnson, N.L. and Kotz, S. (1985a) Some tests for detection of faulty inspection. *Statist. Hefte*, **26**, 19–29. (Ch. 11)

Johnson, N.L. and Kotz, S. (1985b) Some multivariate distributions arising in faulty sampling inspection. In *Multivariate Analysis – VI* (ed. P.R. Krishnaiah), Elsevier, Amsterdam (pp. 311–325). (Ch. 3)

Johnson, N.L. and Kotz, S. (1985c) Some distributions arising as a consequence of errors in inspection. *Naval Res. Logistics Qtly.*, **32**, 35–43. (Ch. 3, 12)

Johnson, N.L. and Kotz, S. (1988a) Estimation from binomial data with classifiers of known and unknown imperfections. *Naval Res. Logistics*, **35**, 147–56. (Ch. 11)

Johnson, N.L. and Kotz, S. (1988b) Effects of errors in inspection on a binary method for isolating nonconforming items. *Austral. J. Statist.*, **30A**, 205–213. (Ch. 10)

Johnson, N.L., Kotz, S. and Rodriguez, R.N. (1984) Inspection errors in link sampling. *Commun. Statist. – Theor. Meth.*, **13**, 1203–1213. (Ch. 4)

Johnson, N.L., Kotz, S. and Rodriguez, R.N. (1985) Statistical effects of imperfect inspection sampling: I. Some basic distributions. *J. Qual. Technol.*, **17**, 1–31. (Ch. 2)

Johnson, N.L., Kotz, S. and Rodriguez, R.N. (1986) Statistical effects of imperfect inspection sampling: II. Double sampling and link sampling. *J. Qual. Technol.*, **18**, 116–138. (Ch. 4)

Johnson, N.L., Kotz, S. and Rodriguez, R.N. (1988) Statistical effects of imperfect inspection sampling: III Screening (group testing). *J. Qual. Technol.*, **20**, 98–124. (Ch. 6, 8)

Johnson, N.L., Kotz, S. and Rodriguez, R.N. (1989) Dorfman–Sterrett (group testing) schemes and the effects of faulty inspection. *Commun. Statist.–Theor. Meth.*, **18**, 1469–1484. (Ch. 3)

Johnson, N.L., Kotz, S. and Rodriguez, R.N. (1990) Statistical effects of imperfect inspection sampling, IV. Modified Dorfman screening procedures. *J. Qual. Technol.*, **22**, 128–137. (Ch. 7, 9, 12)

Johnson, N.L., Kotz, S. and Sorkin, H.L. (1980) Faulty inspection distributions. *Commun. Statist.–Ther. Meth.*, **A9**, 917–922. (Ch. 22)

Johnson, N.L., Kotz, S. and Wang, Q. (1990) Randomized-sequential group testing procedures. *Sequential Analysis.*, **9**, 219–242.

Johnson, N.L., Kotz, S. and Wu, X. (1989) Quota fulfilment and errors in stratification. *Commun. Statist. – Theor. Meth.*, **18**, 1497–1505. (Ch. 12)

Johnson, S.L. and Funke, D.J. (1980) An analysis of human reliability measures in visual inspection. *J. Qual. Technol.*, **12**, 71–74. (Ch. 1)

Jones, H.G. (1933) A note on the n-ages method. *J. Inst. Actu.*, **64**, 318–324. (Ch. 11)

Juran, J.M. (1935) Inspectors' errors in quality control. *Mechanical Engineering*, **57**, 643–644. (Ch. 1)

Juran, J.M. (1968) Operator errors – time for a new look, *Quality Progress*, **1**(2), 9–11. (Ch. 1)

Kemp, A.W. and Kemp, C.D. (1988) A simple inspection scheme for two types of defect. *J. Opl. Res. Soc.*, **39**, 311–315. (Ch. 6)

Kittler, J. and Papu, L.F. (1980) Automatic inspection by lots in presence of classification errors. *Pattern Recognition*, **12**, 237–241. (Ch. 1)

Kotz, S. and Johnson, N.L. (1982a) Effects of false and incomplete identification of defective items on the reliability of acceptance sampling. *Opns. Res.*, **32**, 575–583. (Ch. 2)

Kotz, S. and Johnson, N.L. (1982b) Errors in inspection and grading: distributional aspects of screening and hierarchal screening. *Commun. Statist. – Theor. Meth.*, **11**, 1997–2016. (Ch. 6)

Kotz, S. and Johnson, N.L. (1983) Some distributions arising from faulty inspection with multitype defectives, and applications to grading. *Commun. Statist. – Theor. Meth.*, **12**, 2809–2821. (Ch. 5, 12)

Kotz, S. and Johnson, N.L. (1989) Effects of inspection errors on Dorfman screening procedures with random group size: a note on Kemp and Kemp's 'Simple inspection scheme for two types of defect'. *J. Opl. Res. Soc.*, **40**, 103–107. (Ch. 6)

Kotz, S., Johnson, N.L. and Gunther, N. (1988) Inspection errors in Graff–Roeloffs' modification of hierarchical Dorfman screening procedures. *Commun. Statist. – Theor. Meth.*, **17**, 139–168. (Ch. 9)

Kotz, S., Mao, S. and Johnson, N.L. (1986) Effects of inspection errors on curtailed Dorfman-type procedures. *Commun. Statist. – Theor. Meth.*, **15**, 831–838. (Ch. 8)

Kulapatrapa, S. and Case K.E. (1976) Sequential attribute sampling: inspection error effects and compensation. *Proc. AIIE Systems Eng. Conf.*, 187–196. (Ch. 1)

Kumar, S. and Sobel, M. (1971) Finding a single defective in binomial group-testing. *J. Amer. Statist. Ass.*, **66**, 824–828. (Ch. 10)

Lavin, M. (1946) Inspection efficiency and sampling inspection plans. *J. Amer. Statist. Assoc.*, **41**, 432–438. (Ch. 1)

Lindsay, B.G. (1985) Errors in inspection: integer parameter maximum likelihood in a finite population. *J. Amer. Statist. Assoc.*, **80**, 879–885. (Ch. 11)

Lindsay, B.G. and Roeder, K. (1987) A unified treatment of integer parameter models. *J. Amer. Statist. Assoc.*, **82**, 758–764. (Ch. 11)

Maghsoodloo, S. (1987) Inspection error effects on performance measures of a multistage sampling plan. *AIIE Trans.*, **19**, 340–347. (Ch. 1)

Mao, S., Kotz, S. and Johnson, N.L. (1986) Estimation by moments in a model of faulty inspection. *J. Statist. Comp. Simul.*, **23**, 265–272. (Ch. 11)

McKenzie, R.M. (1958) On the accuracy of inspectors. *Ergonomics*, **1**, 258–272. (Ch. 1)

McKnight, K.A. (1967) *An investigation of the effects of two types of inspector error on sampling inspection plans*, M.S. thesis, Texas Technological College, Lubbock, Texas. (Ch. 1)

Mehravari, N. (1980) Generalized binary binomial group testing for a modified binomial model. *SIAM J. Alg. Discrete Math*, **7**, 159–160. (Ch. 10)

Menzefricke, U. (1984) The effects of variability in inspection error. *J. Qual. Technol.*, **16**, 131–135. (Ch. 1)

Minton, G. (1969) Inspection and correction error in data processing. *J. Amer. Statist. Assoc.*, **64**, 1256–1275. (Ch. 1)

Minton, G. (1972) Verification error in single sampling inspection plans. *J. Amer. Statist. Assoc.*, **67**, 46–54. (Ch. 1)

Moskowitz, H. and Fink, R.K. (1977) A Bayesian algorithm incorporating inspection errors for quality control and auditing, *TIMS Studies in the Management Sciences*, **7**, 79–104, North Holland, Amsterdam. (Ch. 1)

Mundel, A.B. (1984) Group testing. *J. Qual. Technol.*, **16**, 46–54. (Ch. 6)

National Research Council (1989) Statistical models and analysis in auditing. *Statist. Sci.*, **4**, 2–33. (Ch. 1)

Newell, D.J. (1962) Errors in the interpretation of errors in epidemiology. *Amer. J. Publ. Hlth.*, **52**, 1925–1928. (Ch. 1)

Olkin, J., Petkau, A.J. and Zidek, J.V. (1981) A comparison of n estimators for the binomial distribution. *J. Amer. Statist. Assoc.*, **76**, 637–648. (Ch. 11)

Osanaiye, P.A. and Alebiosu, S.A. (1988) Effects of industrial inspection errors on some plans that use the surrounding lot information. *J. Appl. Statist.*, **15**, 295–304. (Ch. 4)

Paramonova, G.G. (1990) Determination of components in linear systems, accounting for errors in identification. (In Russian) *Automatika i Telemekhanika*, **2B**, 152–160. (Ch. 1)

Pfeifer, C.G. and Enis, (1978) Dorfman-type group testing for a modified binomial model. *J. Amer. Statist. Ass.*, **73**, 588–592. (Ch. 8)

Phatak, A.G. (1968) Misclassified data from curtailed sampling plans. *Technometrics*, **10**, 489–496. (Ch. 1)

Rahali, B. and Foote, B.L. (1982) An approach to compensate for uncertainty in knowledge of inspection error. *J. Qual. Technol.*, **14**, 190–195. (Ch. 1)

Rubin, T., Rosenbaum, J. and Cobb, S. (1958) The use of interview data for detection of association in field studies. *J. Chronic Dis.*, **4**, 253–266. (Ch. 1)

Sandelius, M. (1961) On an optimal search procedure. *Amer. Math. Monthly.*, **68**, 133–134. (Ch. 10)

Shewhart, W.A. (1926) Correction of data for errors of measurement. *Bell System Tech. J.*, **5**, 11–26. (Ch. 1)

Shewhart, W.A. (1931) *Economic control of quality of manufactured product.* Van Nostrand, New York. (Ch. 1)

Sinclair, M.A. (1978) A collection of modelling approaches for visual inspection in industry. *Intern. J. Production Res.*, **16**, 275–292. (Ch. 1)

Singh, H.R. (1964) Measurement error in acceptance sampling for attributes. *Indian Soc. Qual. Control Bull.*, **10**, (1/2). (Ch. 1)

Singh, H.R. and Dutta, H.N. (1976) Studies of measurement error in statistical quality control: a survey. *J. Indian Assoc. Productivity*, **1**, 51–56. (Ch. 1)

Smith, J.R. and Duvier, H. (1984) Effect of inspection error on inspection strategies. *Trans. 38th Annual Conf., American Society for Quality Control*, 146–151. (Ch. 1)

Sobel, M. (1968) Group testing to classify efficiently all units in a binomial sample. In *Information and Decision Processes*, (ed. R.E. Machol), McGraw-Hill, New York. (Ch. 10)

Sobel, M. and Groll, P.A. (1959) Group testing to eliminate efficiently all defectives in a binomial sample, *Bell Syst. Tech. J.*, **38**, 1179–1251. (Preface)

Sobel, M. and Groll, P.A. (1966) Binomial group testing with an unknown proportion of defectives. *Technometrics*, **8**, 631–656. (Ch. 10)

Sorkin, H.L. (1978) *The expanded field: confirmation.* Institute of Internal Auditors Inc., Altamonte Springs, Florida. (Ch. 1)

Stephens, K.S. (1979) *How to perform continuous sampling* (ASQC Basic References in Quality Control: Statistical Techniques, **2**) American Society for Quality Control, Milwaukee, Wisconsin. (Ch. 4)

Sterrett, A. (1957) On the detection of defective members of large populations. *Ann. Math. Statist.*, **28**, 1033–1036. (Ch. 7)

Swallow, W.H. (1985) Group testing for estimating infection rates and probabilities of disease transmission. *Phytopathology*, **75**, 882–889. (Ch. 6)

Swallow, W.H. (1987) Relative mean square error and cost

considerations in choosing group size for group testing to estimate infection rates and probabilities of disease transmission. *Phytopathology*, **77**, 1376–1381. (Ch. 6)

Tenenbein, A. (1970) A double sampling scheme for estimating from binomial data with misclassifications. *J. Amer. Statist. Assoc.*, **65**, 1350–1361. (Ch. 11)

Thomas, J., Pasternack, B.S., Vacirca, S.J. and Thompson, D.L. (1973) Applications of group-testing procedures in radiological health. *Health Physics*, **25**, 259–266. (Ch. 6)

Thomas, J., Pasternack, B.S., Bohning, D.E. and Vacirca, S.J. (1975) Methods for reducing exposure to personnel leak-testing sealed radium sources. *Health Physics*, **28**, 111–121. (Ch. 6)

Toms, M. and Patrick, J. (1989) Components of fault finding: symptom interpretation. *Human Factors*, **31**, 465–483. (Ch. 1)

Wallock, D.M. and Adams, K.S. (1970) A comparison of inspector performance measures. *AIIE Trans.*, **2**, 97–105. (Ch. 1)

Watson, G.S. (1961) A study of the group-screening method. *Technometrics*, **3**, 371–388. (Ch. 6)

Wolf, J.K. (1985) Born again group testing: multiaccess communications. *IEEE Trans. Inf. Theory*, **IT31**, 185–191. (Ch. 6)

Wooding, W.M. (1979) A source of bias in attributes testing and a remedy. *J. Qual. Technol.*, **11**, 169–176. (Ch. 1)

Index

Acceptance number 7
  probabilities 44, 45, 46, 47, 51
  sampling procedures 7
    (*see also* sampling plans)
Accounts with errors ix, 5
Accuracy of inspection
  measure of 145
Asymptotic efficiency 167
Average outgoing quality
    (AOQ) 4, 53
AQL 44, 45, 46, 47

Bayes' formula 70
  theorem 56
Bernoulli trials 134
Bias 171
  proportional 164, 165
Binomial distribution 12, 13, 14, 15, 27, 122, 124, 135, 136, 190
  variance of 14, 28
Blood samples x, 78, 195

CSP plans 53, 54, 55
Census data processing 3
Classes 26, 180
Clearing interval 53
Compensation for inspection
    error 5, 8
Compounding of distributions 12, 13, 16, 27, 70
Computer programs 192, 193
Conditional distributions 13, 15, 68
  joint 34
Conditional maximum likelihood
    estimate 157
Confidence intervals 161
Conforming (C) item 6, 11, 33
Conservative test 178
Contaminant in mixture 79
Convolution of binomial
    distributions 50, 113
  of distributions 15, 37, 59, 68
  multifold 27, 72
  of hypergeometric
    distributions 50
  repeated 68
Cost functions 186
Covariance 35, 39, 61, 63, 64, 72, 73, 172, 174, 183
Curtailed group testing
    (*see* titles of Chapters and Sections)

Decision probability 42
Defects in new automobiles 57
Descending factorial 13
Destructive sampling 7
Dorfman procedure
    (*see* titles of Chapters and Sections)
Dorfman-Sterrett group testing
    (*see* titles of Chapters and Sections)

Effective fraction defective 1, 2, 5, 6
EM algorithm 167, 168
Error-free inspection 8, 78, 125, 149, 150, 155, 191
  see also perfect inspection
Estimation (estimator)
  bias of 165, 171

Estimation (*Contd.*)
maximum likelihood 155, 156, 157, 165, 167, 169
moment 155, 158, 163, 167, 169
unbiased 163, 170
Expected cost of sampling 181
Expected length of inspection 54
of production run 91
Expected number of inspected items 52
of tests 80, 83, 84, 99, 114, 130, 136, 137, 139–41, 196
Expected percentage (proportional) reduction 83, 99
Expected savings in testing 78, 79

Factorial moments
ascending 29
conditional 81
descending 13, 16, 17, 27, 28, 37, 38, 82, 159, 175
joint 62, 63, 182
mixed 31, 60, 62
unconditional 81
False indication of nonconformity 80, 175
False negative 96
False positive 80, 175

GEHT 54
Graff and Roeloffs procedure (*see* titles of Chapters and Sections)
Group testing ix, 5, 73 (*see also* titles of Chapters and Sections)

Halving 146, 147
Harmonic interpolation 21
Hypergeometric distribution 12, 13, 99, 103, 122, 124
Hypergeometric-binomial distribution 59
Hierarchal screening 79 (*see* titles of Chapters and Sections)

Imperfect inspection 23, 25
Improper distribution 14
Information matrix 166
Inspection devices 162
strategy 113
Interview method 4
Iteration formulas for expectations 168

Kemp and Kemp procedure 91-6
Kronecker product 173

Likelihood function 156, 165
Log likelihood 156
conditional 156
Medical applications x, 78, 143
Medical examinations 4
MIL-STD-105D 46
Minimum expected cost 7
Moment ratios 176, 177
Multi-access communication theory 87
Multi-hypergeometric distribution 27, 36
Multi-hypergeometric multinomial distribution 170, 171, 182, 184
Multinomial coefficient 27, 33
distribution 68, 163, 171, 182
Multiple sampling 43
Multistage procedures 6, 105 (*see also* Multiple sampling)
Multivariate hypergeometric-convoluted binomial distribution 37, 60
Multivariate hypergeometric distribution 34, 52, 59, 60, 64, 70, 73, 182

NC pattern 66, 67, 169, 183
Negative hypergeometric distribution 29
Neighboring lots 49
Nonconforming (NC) items 5, 6, 11, 33
Nonconformity pattern 57 (*see also* NC pattern)

Normalized value (norm) 29
Nuisance parameter 178

Optical sampling plan 5
Overall acceptance probability 43, 48

Percentage of misclassification 3
Perfect inspection 45
Positive result in group testing 80, 83, 84
Potentially declared NC items 118
Power function 178, 179
Prevalence rate 78, 196 (*see also* Proportion of NC items)
Probabilities of errors 6, 7
Probability of correct identification 80, 83, 84, 99, 105, 106, 114, 121, 123, 131, 132, 137, 139–41
Proportion of NC items 156, 157
Proportion of nonconforming 17, 18, 19, 20

Quality control systems 4

Randomized sequential procedure 98, 112 (*see also* titles of Chapters and Sections)
Rectification inspection schemes 4
Regression function 39, 40
Repeated convolution 68
Repeated observations 8, 74

S2-L2 plan 4
S3-L3 plan 4
Sampling fraction 53, 175
  inspection 1
  plan 7
  with replacement 11, 38
  without repalcement 11, 21, 26, 49
Sequential sampling inspection 4
Single sampling schemes 2, 3, 7, 11, 43
  Bayesian 5
Skip-lot plan 5
Split of sequence 146
Statistical differences 164
Strata 26, 181 (*see also* Classes)
Subgroups 86, 88, 137, 138
Sufficient statistic 41, 156
Syphilitic antigen 78, 198

True fraction defective 1, 2 (*see also* Proportion of nonconfroming)
Two-by-two tables 3, 4

Uniform inspection accuracy 172
Uninspected pattern 72
Urine specimen 77
US Armed Forces 78
US Bureau of the Census 4

Value of excess 186, 187
Variance–covariance matrix 164, 172
  asymptotic 166
Variances
  asymptotic 160

Waiting time distribution 14, 28, 30, 31